T0309563

Applied Mathematical Sciences
Volume 87

Applied Mathematical Sciences

1. *John:* Partial Differential Equations, 4th ed.
2. *Sirovich:* Techniques of Asymptotic Analysis.
3. *Hale:* Theory of Functional Differential Equations, 2nd ed.
4. *Percus:* Combinatorial Methods.
5. *von Mises/Friedrichs:* Fluid Dynamics.
6. *Freiberger/Grenander:* A Short Course in Computational Probability and Statistics.
7. *Pipkin:* Lectures on Viscoelasticity Theory.
8. *Giacaglia:* Perturbation Methods in Non-Linear Systems.
9. *Friedrichs:* Spectral Theory of Operators in Hilbert Space.
10. *Stroud:* Numerical Quadrature and Solution of Ordinary Differential Equations.
11. *Wolovich:* Linear Multivariable Systems.
12. *Berkovitz:* Optimal Control Theory.
13. *Bluman/Cole:* Similarity Methods for Differential Equations.
14. *Yoshizawa:* Stability Theory and the Existence of Periodic Solution and Almost Periodic Solutions.
15. *Braun:* Differential Equations and Their Applications, 3rd ed.
16. *Lefschetz:* Applications of Algebraic Topology.
17. *Collatz/Wetterling:* Optimization Problems.
18. *Grenander:* Pattern Synthesis: Lectures in Pattern Theory, Vol I.
19. *Marsden/McCracken:* Hopf Bifurcation and Its Applications.
20. *Driver:* Ordinary and Delay Differential Equations.
21. *Courant/Friedrichs:* Supersonic Flow and Shock Waves.
22. *Rouche/Habets/Laloy:* Stability Theory by Liapunov's Direct Method.
23. *Lamperti:* Stochastic Processes: A Survey of the Mathematical Theory.
24. *Grenander:* Pattern Analysis: Lectures in Pattern Theory, Vol. II.
25. *Davies:* Integral Transforms and Their Applications, 2nd ed.
26. *Kushner/Clark:* Stochastic Approximation Methods for Constrained and Unconstrained Systems.
27. *de Boor:* A Practical Guide to Splines.
28. *Keilson:* Markov Chain Models—Rarity and Exponentiality.
29. *de Veubeke:* A Course in Elasticity.
30. *Sniatycki:* Geometric Quantization and Quantum Mechanics.
31. *Reid:* Sturmian Theory for Ordinary Differential Equations.
32. *Meis/Markowitz:* Numerical Solution of Partial Differential Equations.
33. *Grenander:* Regular Structures: Lectures in Pattern Theory, Vol. III.
34. *Kevorkian/Cole:* Perturbation methods in Applied Mathematics.
35. *Carr:* Applications of Centre Manifold Theory.
36. *Bengtsson/Ghil/Källén:* Dynamic Meteorology: Data Assimilation Methods.
37. *Saperstone:* Semidynamical Systems in Infinite Dimensional Spaces.
38. *Lichtenberg/Lieberman:* Regular and Stochastic Motion.
39. *Piccini/Stampacchia/Vidossich:* Ordinary Differential Equations in R^n.
40. *Naylor/Sell:* Linear Operator Theory in Engineering and Science.
41. *Sparrow:* The Lorenz Equations: Bifurcations, Chaos, and Strange Attractors.
42. *Guckenheimer/Holmes:* Nonlinear Oscillations, Dynamical Systems and Bifurcations of Vector Fields.
43. *Ockendon/Tayler:* Inviscid Fluid Flows.
44. *Pazy:* Semigroups of Linear Operators and Applications to Partial Differential Equations.
45. *Glashoff/Gustafson:* Linear Optimization and Approximation: An Introduction to the Theoretical Analysis and Numerical Treatment of Semi-Infinite Programs.
46. *Wilcox:* Scattering Theory for Diffraction Gratings.
47. *Hale et al.:* An Introduction to Infinite Dimensional Dynamical Systems—Geometric Theory.
48. *Murray:* Asymptotic Analysis.
49. *Ladyzhenskaya:* The Boundary-Value Problems of Mathematical Physics.
50. *Wilcox:* Sound Propagation in Stratified Fluids.
51. *Golubitsky/Schaeffer:* Bifurcation and Groups in Bifurcation Theory, Vol. I.

(continued following index)

Ricardo Weder

Spectral and Scattering Theory for Wave Propagation in Perturbed Stratified Media

Springer-Verlag
New York Berlin Heidelberg London
Paris Tokyo Hong Kong Barcelona

Ricardo Weder
Instituto de Investigaciones en Matemáticas Aplicadas y en Sistemas
Universidad Nacional Autónoma de México
México, D.F., 01000
México

Weder, Ricardo.
Spectral and scattering theory for wave propagation in perturbed stratified media / Ricardo Weder.
p. cm. — (Applied mathematical sciences)
Includes bibliographical references.
1. Wave-motion, Theory of. 2. Spectral theory (Mathematics) 3. Scattering (Physics) I. Title. II. Series: Applied mathematical sciences (Springer-Verlag New York Inc.)
QA1.A627
[QA927]
510—dc20
[530.1'24] 90-9845

Printed on acid-free paper.

Photocomposed pages prepared from author's TEX file.
Printed and bound by R.R. Donnelley & Sons, Harrisonburg, VA.
Printed in the United States of America.

9 8 7 6 5 4 3 2 1

ISBN 0-387-97357-5 Springer-Verlag New York Berlin Heidelberg
ISBN 3-540-97357-5 Springer-Verlag Berlin Heidelberg New York

To my wife Teresa, and my children
Natalie Danitza and Ricardo Eugenio

Contents

1
Introduction

The propagation of acoustic and electromagnetic waves in stratified media is a subject that has profound implications in many areas of applied physics and in engineering, just to mention a few, in ocean acoustics, integrated optics, and wave guides. See for example Tolstoy and Clay 1966, Marcuse 1974, and Brekhovskikh 1980.

As is well known, stratified media, that is to say media whose physical properties depend on a single coordinate, can produce guided waves that propagate in directions orthogonal to that of stratification, in addition to the free waves that propagate as in homogeneous media.

When the stratified media are perturbed, that is to say when locally the physical properties of the media depend upon all of the coordinates, the free and guided waves are no longer solutions to the appropriate wave equations, and this leads to a rich pattern of wave propagation that involves the scattering of the free and guided waves among each other, and with the perturbation. These phenomena have many implications in applied physics and engineering, such as in the transmission and reflexion of guided waves by the perturbation, interference between guided waves, and energy losses in open wave guides due to radiation.

The subject matter of this monograph is the study of these phenomena.

We present here for the first time a complete and self-contained study of acoustic and electromagnetic wave propagation in perturbed stratified media from the point of view of the modern constructive stationary spectral and scattering theory. Many results are published here for the first time (see Notes).

In Chapter 2 we study the case of propagation of acoustic waves in a perturbed Pekeris velocity profile. This problem exhibits the main features of wave propagation in perturbed stratified media and is simple enough to present the powerful techniques of modern constructive spectral and scattering theory without the technicalities that are required to study more complicated models and make the proofs more difficult to follow.

In Chapter 3 we study the case of electromagnetic wave propagation in three dimensional dielectric wave guides, described by the vector Maxwell system of equations.

The intended readers of this monograph are applied mathematicians, physicists, and engineers working in wave propagation in stratified media, as well as mathematicians and mathematical physicists interested in spectral and scattering theory.

This monograph provides the researcher and graduate student interested in wave propagation in stratified media with a rigorous framework in order to understand from first principles the problems involved in the subject as well as the methods of solution.

To the reader expert in spectral and scattering theory this monograph provides results in a new field of applications, with new methods of solutions. One example of such a new method presented here is the proof of the absence of positive eigenvalues of the perturbed acoustic propagator in Theorem 5.1 in Chapter 2.

The monograph is organized as follows.

In Chapter 2 we consider the case of acoustic waves.

In Chapter 3 we study the case of electromagnetic waves.

In the Appendices 1 and 2 we present results that are used in the previous chapters.

The bibliographical notes are on page 179.

The references are listed at the end of the monograph in alphabetic order.

Mexico City, August 1989

2

Propagation of Acoustic Waves

§1. The Unperturbed Acoustic Propagator

The propagation of acoustic waves in a $n+1$ dimensional stratified fluid is described by the following wave equation

$$\frac{\partial^2}{\partial_t^2} u(x,y,t) - c_0^2(y)\Delta u(x,y,t) = 0, \tag{1.1}$$

where Δ denotes the $n+1$ dimensional Laplacian

$$\Delta = \sum_{i=1}^{n} \frac{\partial^2}{\partial x_i^2} + \frac{\partial^2}{\partial_y^2}, \tag{1.2}$$

where the derivatives are in distribution sense $x = (x_1, x_2, \cdots x_n) \in \mathbf{R}^n$, $y \in \mathbf{R}$, $t \in \mathbf{R}$.

The acoustic potential $u(x,y,t)$ is a real valued function of x, y and t. $c_0(y)$ is a real valued measurable function on $\mathbf{R}$ that is bounded below and above

$$0 < c_m \leq c_0(y) \leq c_M < \infty, \tag{1.3}$$

for almost every y and some positive constants c_m and c_M.

The function $c_0(y)$ represents the speed of propagation of sound waves in the fluid.

We denote by $L^2(\mathbf{R}^{n+1})$ the Hilbert space of all complex valued Lebesgue square integrable functions on $\mathbf{R}^{n+1}$ with scalar product

$$(\varphi, \psi)_{L^2(\mathbf{R}^{n+1})} = \int \varphi(x,y)\, \overline{\psi}(x,y)\, dxdy, \tag{1.4}$$

for $\varphi,\ \psi \in L^2(\mathbf{R}^{n+1})$.

For m any nonnegative integer we denote by $H_m(\mathbf{R}^{n+1})$ the Sobolev space of order m consisting of all complex valued Lebesgue square integrable functions on $\mathbf{R}^{n+1}$ whose distributional derivatives of order up to m are also square integrable, with scalar product

$$(\varphi, \psi)_{H_m(\mathbf{R}^{n+1})} = \sum_{|\alpha| \le m} (D^\alpha \varphi,\ D^\alpha\ \psi)_{L^2(\mathbf{R}^{n+1})}, \tag{1.5}$$

for $\varphi,\ \psi \in H_m(\mathbf{R}^{n+1})$, where we use the standard multiindex notation

$$D^\alpha \ = \ \frac{\partial^{\alpha_1}}{\partial_{x_1}{}^{\alpha_1}}\ \ \frac{\partial^{\alpha_2}}{\partial_{x_2}{}^{\alpha_2}} \cdots \frac{\partial^{\alpha_n}}{\partial_{x_n}{}^{\alpha_n}}\ \ \frac{\partial^{\alpha_{n+1}}}{\partial_{y}{}^{\alpha_{n+1}}}, \tag{1.6}$$

for $\alpha = (\alpha_1, \alpha_2, \cdots, \alpha_{n+1})$, and where $|\alpha| = \sum_{j=1}^{n+1} \alpha_j$.

We denote also by Δ the selfadjoint realization of Δ in $L^2(\mathbf{R}^{n+1})$ with domain $H_2(\mathbf{R}^{n+1})$.

We denote by $\mathcal{H}_0$ the Hilbert space of all complex valued Lebesgue square integrable functions on $\mathbf{R}^{n+1}$ with scalar product

$$(\varphi,\ \psi)_{\mathcal{H}_0} = \int \varphi(x,y)\ \overline{\psi}(x,y)\ c_0^{-2}(y)\ dxdy, \tag{1.7}$$

for $\varphi,\ \psi \in \mathcal{H}_0$. Note that because of (1.3) the norms of $\mathcal{H}_0$ and of $L^2(\mathbf{R}^{n+1})$ are equivalent.

The unperturbed acoustic propagator is the following operator in $\mathcal{H}_0$

$$A_0\ \varphi = -c_0^2(y)\ \Delta\varphi, \tag{1.8}$$

with domain

$$D(A_0) = H_2(\mathbf{R}^{n+1}). \tag{1.9}$$

A_0 is a selfadjoint and positive operator in $\mathcal{H}_0$ (this follows easily from the fact that $-\Delta$ is selfadjoint and positive in $L^2(\mathbf{R}^{n+1})$).

The spectral analysis and the eigenfunctions expansion theory of A_0 have been thoroughly studied in Wilcox 1984, for the class of profiles $c_0(y)$ that satisfy the following condition

$$\pm \int_0^{\pm\infty} |c_0(y) - c_\pm|\ dy\ < \infty, \tag{1.10}$$

for some positive constants c_+, and c_-. We state below some results from Wilcox 1984 that we will need later.

We denote by $\mathcal{F}$ the Fourier transform on the x variables as an unitary operator on $\mathcal{H}_0$

$$(\mathcal{F}\ \varphi)(k,y)\ =\ s - \lim_{M\to\infty} \frac{1}{(2\pi)^{n/2}} \int_{|x|\le M} e^{-ik\cdot x}\ \varphi(x,y)\ dx, \tag{1.11}$$

where the limit exists in the strong topology in $\mathcal{H}_0$.

We denote by $\mathcal{L}$ the Hilbert space of complex valued Lebesgue square integrable functions on $\mathbf{R}$ with the scalar product

$$(\varphi,\ \psi)_{\mathcal{L}} = \int \varphi(y)\ \overline{\psi}(y) c_0^{-2}(y)\ dy. \tag{1.12}$$

We have that

$$A_0 = \mathcal{F}^{-1} \hat{A}_0\ \mathcal{F}, \tag{1.13}$$

where

$$\hat{A}_0 = \oplus \int_{\mathbf{R}^n} A(k)\ dk, \tag{1.14}$$

where $A(k),\ k \in \mathbf{R}^n$ is the following selfadjoint operator in $\mathcal{L}$

$$A(k)\varphi = c_0^2(y) \left[-\frac{d^2}{dy^2} + k^2 \right] \varphi, \tag{1.15}$$

with domain

$$D(A(k)) = \{\varphi \in \mathcal{L} :\ \frac{d^2}{dy^2}\ \varphi \in \mathcal{L}\}. \qquad (1.16).$$

Note that

$$\mathcal{H}_0 = \oplus \int_{\mathbf{R}^n} \mathcal{L}\ dx. \tag{1.17}$$

The spectral analysis of A_0 is thus reduced to that one of $A(k)$. For the definition of the direct integrals in (1.14) and (1.17), see Dixmier 1969, Chapter II.

We give below some well known definitions concerning the spectrum that we will frequently use (see Kato 1976).

For X, Y any pair of Banach spaces we denote by $B(X, Y)$ the Banach space of all bounded linear operators from X into Y. When $X = Y$ we use the notation $B(X)$ instead of $B(X, X)$.

Let B be any closed linear operator on a Banach space X. We denote by $\rho(B)$ the resolvent set of B. Namely $\rho(B)$ is the set of all $z \in \mathbf{C}$ such that $B - z$ is one to one with $(B - z)^{-1} \in B(X)$ · $\rho(B)$ is an open set. For $z \in \rho(B)$ we denote by

$$R(z) = (B - z)^{-1}, \tag{1.18}$$

the resolvent of B. $R(z)$ is an analytic function of $z \in \rho(B)$ with values on $B(X)$. We denote by $\sigma(B) = \mathbf{C} \setminus \rho(B)$, the spectrum of B. $\sigma(B)$ is a closed set.

Suppose that B is a selfadjoint operator on a Hilbert space $\mathbf{H}$.

We denote by $\sigma_d(B)$ the discrete spectrum of B. Namely $\sigma_d(B)$ is the set of eigenvalues of B that are isolated points in the spectrum, and have finite multiplicity. We designate by $\sigma_e(B) = \sigma(B) \setminus \sigma_d(B)$ the essential spectrum of B.

By $E(\lambda)$, $\lambda \in \mathbf{R}$, we denote the family of spectral projectors associated with the selfadjoint operator B (see Kato 1976, Section 5 of Chapter 6).

For any $\varphi \in \mathbf{H}$ we denote by

$$m_\varphi(\Delta) = (E(\Delta)\varphi, \varphi), \tag{1.19}$$

for Δ any Borel set in $\mathbf{R}$, the spectral measure associated with φ. We say (see Kato 1976, Section 1 of Chapter 10) that φ is absolutely continuous if the measure (1.19) is absolutely continuous with respect to the Lebesgue measure, and we denote by $\mathcal{H}_{ac}(B)$ the closure of the set of all finite linear combinations of absolutely continuous vectors in $\mathbf{H}$.

Similarly φ is singular continuous if the measure (1.19) is singular continuous with respect to the Lebesgue measure. By $\mathcal{H}_{sc}(B)$ we denote the closure of the set of all finite linear combinations of singular continuous vectors. By $\mathcal{H}_{pp}(B)$ we denote the closure of the set of all finite linear combinations of all the eigenvectors of B.

It is known (see Kato 1976, Chapter 10) that

$$\mathbf{H} = \mathcal{H}_{pp}(B) \oplus \mathcal{H}_{ac}(B) \oplus \mathcal{H}_{sc}(B), \tag{1.20}$$

and that B is reduced by the decomposition in (1.20)

$$B = B_{pp} \oplus B_{ac} \oplus B_{sc}. \tag{1.21}$$

$\mathcal{H}_{pp}(B)$, $\mathcal{H}_{ac}(B)$, and $\mathcal{H}_{sc}(B)$ are respectively the pure point, absolutely continuous, and singular continuous subspaces of B, and B_{pp}, B_{ac}, and B_{sc} are respectively the pure point, absolute continuous, and singular continuous parts of B.

We say that B has no singular continuous spectrum if $\mathcal{H}_{sc}(B) = \{0\}$, and we say that B is absolutely continuous if $\mathcal{H}_{pp}(B) = \mathcal{H}_{sc}(B) = \{0\}$.

We denote by $\sigma_c(B)$ the continuous spectrum of B. Namely $\sigma_c(B) = \sigma(B_{ac} \oplus B_{sc})$. By $\sigma_p(B)$ we denote the set of all the eigenvalues of B.

In Wilcox 1984, Chapter 3, it is proven that $\sigma_p(A(k)) \subset [c_m^2 k^2,\ c_+^2 k^2]$, and that $\sigma_c(A(k)) = [c_+^2 k^2,\ \infty)$. Moreover, eigenfunctions expansion theorems are proven for $A(k)$, $k \neq 0$, in terms of the eigenfunctions corresponding to the eigenvalues of $A(k)$, and generalized eigenfunctions $\psi_0(y, |k|, \lambda)$, for $c_+^2 k^2 < \lambda < c_-^2 k^2$, and $\psi_\pm(y, |k|, \lambda)$, for $\lambda > c_-^2 k^2$. The spectrum in $(c_+^2 k^2, c_-^2 k^2)$ is of multiplicity one, and the spectrum on $(c_+^2 k^2, \infty)$ is of multiplicity two. When $c_+ = c_-$, $\psi_0(y, |k|, \lambda)$ is absent.

Furthermore a generalized eigenfunctions expansion theory for A_0 is obtained from the one for $A(k)$.

From the generalized eigenfunctions expansions a spectral representation of A_0 is constructed.

The spectral representation of A_0 is the starting point in the proof of the limiting absorption principle that we give in section 2.

The validity of the limiting absorption principle requires some regularity of the spectral measure of A_0. This in turn implies (via the trace maps constructed in section 2) regularity properties of the normalized eigenfunctions and generalized eigenfunctions of $A(k)$. The derivation of these regularity properties is a problem in ordinary differential equations that requires techniques different from those used to obtain the main stream of the results in this monograph, namely the spectral and scattering theory of partial differential operators.

In order to give a unified presentation and avoid being burdened by technicalities, we have chosen to consider the main example of a velocity profile, namely the slab

$$c_0(y) = \begin{cases} c_+, & y \geq h, \\ c_h, & 0 \leq y < h, \\ c_-, & y < 0, \end{cases} \tag{1.22}$$

where c_+, c_h, c_-, and h are positive constants. This corresponds to three slabs of fluids respectively in the regions $y \geq h$, $0 \leq y < h$, and $y < 0$, characterized by the parameters c_+, c_h, and c_-. We will assume that $c_+ \leq c_-$ and that $c_h < c_+$. In this case the region $0 \leq y \leq h$ is able to guide an infinite number of waves (guided modes). In the case $c_h \geq c_+$ there are no guided modes, and the problem is simpler, but less interesting from the point of view of the applications.

The Pekeris profile (1.22) was first studied in the modern literature in Pekeris 1948, and is the standard model to describe guided waves in a stratified fluid.

In the case of the Pekeris profile the eigenfunctions and generalized eigenfunctions are explicitly evaluated in terms of elementary functions. This allows us to prove that the trace maps (and the spectral measure) have the required regularity in a simple way.

Once the trace maps are proven to exist and to be locally Hölder continuous, the methods in spectral and scattering theory that we use apply. In fact the results of this monograph are true for any profile that allows for the construction of trace maps that are locally Hölder continuous.

In the case of the Pekeris profile the operator $A(k)$, $k \neq 0$, has a finite number of eigenvalues of multiplicity one. There are numbers ρ_j, $j = 1, 2, 3, \cdots$ such that $\rho_1 \geq 0$, $\rho_{j+1} > \rho_j$, $j = 1, 2, \cdots$, and $\lim_{j \to \infty} \rho_j = \infty$, and functions $\lambda_j(\rho)$ from $U_j = (\rho_j, \infty)$ to $\mathbf{R}^+$ such that for $\rho_j < |k| \leq \rho_{j+1}$, $A(k)$ has exactly j eigenvalues of multiplicity one given by $\lambda_1(|k|) < \lambda_2(|k|) < \cdots < \lambda_j(|k|)$. These functions have the following properties expressed in terms of $w_j(\rho) = \sqrt{\lambda_j(\rho)}$.

1.

$$c_h \rho < w_j(\rho) < c_+ \rho, \quad \rho \in U_j, \tag{1.23}$$

and

$$\lim_{\rho \to \rho_j} w_j(\rho) = c_+ \rho_j, \ \lim_{\rho \to \infty} \frac{1}{\rho} w_j(\rho) = c_h. \tag{1.24}$$

2.

$w_j(\rho)$ is analytic for $\rho \in [\rho_j, \infty)$ when it is defined at $\rho = \rho_j$ by $w_j(\rho_j) = c_+ \rho_j$. Moreover,

$$\frac{d}{d\rho} w_j(\rho) > 0, \text{ for } \rho \in [\rho_j, \infty), \tag{1.25}$$

and

$$\frac{d}{d\rho} w_j(\rho)|_{\rho=\rho_j} = c_+. \tag{1.26}$$

For these properties see Appendix 1, where also explicit formulas for the eigenfunctions and generalized eigenfunctions are given. In Appendix 1, we follow Wilcox 1976, and 1984.

Let $\psi_j(y, |k|)$, $j = 1, 2, 3$ be the eigenfunction of $A(k)$ corresponding to the eigenvalue $\lambda_j(|k|)$, normalized to one, and let $\psi_0(y, |k|, \lambda)$ and $\psi_\pm(y, |k|, \lambda)$ be the generalized eigenfunctions, constructed in Appendix 1.

In Wilcox 1984, Chapter 3, a theorem in generalized eigenfunctions expansion for A_0 is proven in terms of the following generalized eigenfunctions:

$$\psi_\pm(x, y, k, \lambda) = \frac{1}{(2\pi)^{n/2}} e^{ik \cdot x} \psi_\pm(y, |k|, \lambda), \tag{1.27}$$

for $(k, \lambda) \in \Omega$,

$$\psi_0(x, y, k, \lambda) = \frac{1}{(2\pi)^{n/2}} e^{ik \cdot x} \psi_0(y, |k|, \lambda), \tag{1.28}$$

for $(k, \lambda) \in \Omega_0$, and

$$\psi_j(x, y, k) = \frac{1}{(2\pi)^{n/2}} e^{ik \cdot x} \psi_j(y, |k|), \tag{1.29}$$

for $k \in \Omega_j$, $\quad j = 1, 2, 3 \cdots$, and where

$$\Omega = \{(k, \lambda) \in \mathbf{R}^{n+1} \ : \ k \neq 0, \ c_-^2 k^2 < \lambda\}, \tag{1.30}$$

$$\Omega_0 = \{(k, \lambda) \in \mathbf{R}^{n+1} \ : \ k \neq 0, \ c_+^2 k^2 < \lambda < c_-^2 k^2\}, \tag{1.31}$$

and for $j = 1, 2, 3 \cdots$

$$\Omega_j = \{k \in \mathbf{R}^n \setminus 0 \ : \ |k| \in U_j\}. \tag{1.32}$$

If $c_+ = c_-$, $\psi_0(x, y, k, \lambda)$ is not defined.

We find it convenient to reformulate Wilcox's eigenfunctions expansion theorem in terms of appropriate polar coordinates, in order to express the spectral representation of A_0 in a form that is suitable for the proof of the limiting absorption principle.

We denote by S_r^n the sphere of center zero and radius r in $\mathbf{R}^{n+1}$. We define

$$S_c = S_+ \cup S_0 \cup S_-, \tag{1.33}$$

where

$$S_+ = \{w = (w_0, \overline{w}) \in S^n_{\frac{1}{c_+}} \ : \ w_0 \ > \ a|\overline{w}|\}, \tag{1.34}$$

$$S_0 = \{w = (w_0, \overline{w}) \in S^n_{\frac{1}{c_+}} \ : \ 0 \ < \ w_0 \ < \ a|\overline{w}|\}, \tag{1.35}$$

$$S_- = \{w = (w_0, \overline{w}) \in S^n_{\frac{1}{c_-}} \ : \ w_0 \ < \ 0\}, \tag{1.36}$$

where $a = \left(\frac{c_-^2}{c_+^2} - 1\right)^{1/2}$, and $|\overline{w}| = (w_1^2 + \cdots + w_n^2)^{1/2}$.

We denote by χ_+ the following analytic bijection from Ω onto $(0, \infty) \times S_+$,

$$\chi_+(k, \lambda) = (\lambda, w_+(k, \lambda)), \tag{1.37}$$

where

$$w_+(k, \lambda) = (\gamma_+, \overline{w}), \tag{1.38}$$

and

$$\gamma_+ = \left(\frac{1}{c_+^2} - |\overline{w}|^2\right)^{1/2}, \quad \overline{w} = \frac{k}{\sqrt{\lambda}}. \tag{1.39}$$

Similarly we denote by χ_0 the following analytic bijection from Ω_0 onto $(0, \infty) \times S_0$,

$$\chi_0(k, \lambda) = (\lambda, w_+(k, \lambda)), \tag{1.40}$$

with $w_+(k, \lambda)$ as in (1.38), (1.39).

Finally we denote by χ_- the following analytic bijection from Ω onto $(0, \infty) \times S_-$,

$$\chi_-(k, \lambda) = (\lambda, w_-(k, \lambda)), \tag{1.41}$$

where

$$w_-(k, \lambda) = (-\gamma_-, \overline{w}), \tag{1.42}$$

and

$$\gamma_- = \left(\frac{1}{c_-^2} - |\overline{w}|^2 \right)^{1/2}, \quad \overline{w} = \frac{k}{\sqrt{\lambda}}. \tag{1.43}$$

We define the following composite generalized eigenfunction

$$\phi_0(x, y, \lambda, w) = c_+^{\frac{1}{2}} \lambda^{\frac{n}{4}} \gamma_+^{\frac{1}{2}} \psi_+(x, y, k, \lambda), \tag{1.44}$$

where $(k, \lambda) = \chi_+^{-1}(\lambda, w)$, for $(\lambda, w) \in (0, \infty) \times S_+$,

$$\phi_0(x, y, \lambda, w) = c_+^{\frac{1}{2}} \lambda^{\frac{n}{4}} \gamma_+^{\frac{1}{2}} \psi_0(x, y, k, \lambda), \tag{1.45}$$

where $(k, \lambda) = \chi_0^{-1}(\lambda, w)$, for $(\lambda, w) \in (0, \infty) \times S_0$,

$$\phi_0(x, y, \lambda, w) = c_-^{\frac{1}{2}} \lambda^{\frac{n}{4}} \gamma_-^{\frac{1}{2}} \psi_-(x, y, k, \lambda), \tag{1.46}$$

where $(k, \lambda) = \chi_-^{-1}(\lambda, w)$, for $(\lambda, w) \in (0, \infty) \times S_-$. Notice that we have multiplied the functions $\psi_\pm(x, y, k, \lambda)$ and $\psi_0(x, y, k, \lambda)$ by the square root of the Jacobian corresponding to the transformations $\chi_\pm^{-1}$ and χ_0^{-1}.

We find it convenient to parametrize the generalized eigenfunctions (1.29) in terms of $\lambda_j(|k|)$. In order to do so we denote

$$\lambda_j = c_+^2 \rho_j^2, \quad j = 1, 2, 3, \cdots \tag{1.47}$$

Moreover notice that since $\frac{d}{d\rho} \lambda_j(\rho) > 0$, $\lambda_j(\rho)$ is a bijection from $\overline{U}_j$ onto $\overline{0}_j$, where $0_j = (\lambda_j, \infty)$, and where $\overline{U}_j$, $\overline{0}_j$, denote respectively the closure of U_j, and 0_j.

Let $\beta_j(\lambda)$ be the inverse of $\lambda_j(\rho)$, that is to say $\beta_j(\lambda)$ is a function from $\overline{0}_j$ onto $\overline{U}_j$ and

$$\beta_j(\lambda) = \rho \iff \lambda = \lambda_j(\rho), \tag{1.48}$$

for $\lambda \in \overline{0}_j, \quad \rho \in \overline{U}_j$.

We define for $\lambda \in 0_j$, $\nu \in S_1^{n-1}$

$$\phi_j(x, y, \lambda, \nu) = (\beta_j(\lambda))^{\frac{n-1}{2}} (\beta_j'(\lambda))^{\frac{1}{2}} \psi_j(x, y, \beta_j(\lambda)\nu). \tag{1.49}$$

We denote respectively by $L^2(S_c)$ and $L^2(S_1^{n-1})$ the Hilbert space of measurable functions on S_c and the Hilbert space of measurable functions S_1^{n-1} that are square integrable with respect to the measure induced on S_c and S_1^{n-1} by the Lebesgue measure on $\mathbf{R}^{n+1}$ and $\mathbf{R}^n$, with scalar product

$$(\varphi, \psi)_{L^2(S_c)} = \int_{S_c} \varphi(\omega) \overline{\psi}(\omega) \, d\omega, \tag{1.50}$$

$$(\varphi, \psi)_{L^2(S_1^{n-1})} = \int_{S_1^{n-1}} \varphi(\nu) \overline{\psi}(\nu) \, d\nu, \tag{1.51}$$

where $d\omega$ and $d\nu$ denote respectively the measures on S_c and S_1^{n-1}.

We denote by

$$\hat{\mathcal{H}}_0 = L^2((0,\infty),\ L^2(S_c)), \tag{1.52}$$

the Hilbert space of measurable functions on $(0,\infty)$ with values on $L^2(S_c)$ that are square integrable with respect to the Lebesgue measure on $(0,\infty)$, with scalar product

$$(\varphi,\psi)_{\hat{\mathcal{H}}_0} = \int_0^\infty d\rho(\varphi(\rho),\ \psi(\rho))_{L^2(S_c)}. \tag{1.53}$$

The spaces

$$\hat{\mathcal{H}}_j = L^2(0_j,\ L^2(S_1^{n-1})), \tag{1.54}$$

$j = 1,2,3,\cdots$, are similarly defined. We also introduce

$$\hat{\mathcal{H}} = \bigoplus_{j=0}^{\infty} \hat{\mathcal{H}}_j. \tag{1.55}$$

The following theorem is a reformulation of results of Wilcox 1984, Chapter 3.

Theorem 1.1

For every $\varphi \in \mathcal{H}_0$ the following limits

$$\hat{\varphi}_j(\lambda,\omega) = s - \lim_{M\to\infty} \int_{\mathbf{R}_M^{n+1}} \overline{\phi}_j(x,y,\lambda,\omega)\ \varphi(x,y)\ c_0^{-2}(y)\ dxdy, \tag{1.56}$$

$j = 0,1,2,\cdots$, where $R_M^{n+1} = \{(x,y) \in \mathbf{R}^{n+1}\ :\ |x| + |y| \le M\}$, exist on the strong topology in $\hat{\mathcal{H}}_j$, and where $\omega \in S_c$ for $j = 0$, $\omega \in S_1^{n-1}$ for $j = 1,2,3,\cdots$.

The operators

$$(F_j\varphi)(\lambda,\omega) = \hat{\varphi}_j(\lambda,\omega), \tag{1.57}$$

$j = 0,1,2,3,\cdots$, are partially isometric from $\mathcal{H}_0$ onto $\hat{\mathcal{H}}_j$. The adjoint operators are given by

$$(F_0^* f)(x,y) = s - \lim_{M\to\infty} \int_{(0<\lambda<M)\times S_c} \phi_0(x,y,\lambda,\omega) f(\lambda,\omega)\ d\lambda d\omega, \tag{1.58}$$

$$(F_j^* f)(x,y) = s - \lim_{M\to\infty} \int_{(\lambda_j < \lambda < M)\times S_1^{n-1}} \phi_j(x,y,\lambda,\nu) f(\lambda,\nu)\, d\lambda d\nu, \tag{1.59}$$

$j = 1, 2, 3, \cdots$, where the limits exist on the strong topology on $\mathcal{H}_0$.

Let F be the operator from $\mathcal{H}_0$ onto $\hat{\mathcal{H}}$ given for $\varphi \in \mathcal{H}_0$ by

$$F\varphi = (F_0\varphi,\ F_1\varphi,\ F_2\varphi, \cdots). \tag{1.60}$$

Then F is an unitary operator from $\mathcal{H}_0$ onto $\hat{\mathcal{H}}$ and for every $\varphi \in \mathcal{H}_0$

$$F\ A_0\varphi = (\lambda\ F_0\varphi,\ \lambda\ F_1\varphi,\ \lambda\ F_2\varphi, \cdots). \tag{1.61}$$

Proof. This theorem is a reformulation in polar coordinates of the results in Wilcox 1984, Chapter 3 §8, and §9.

Q. E. D.

For $\lambda_j < \lambda \le \lambda_{j+1}, \quad j = 1, 2, 3, \cdots$ we define

$$\hat{\mathcal{H}}(\lambda) = L^2(S_c) \bigoplus_{k=1}^{j} L^2(S_1^{n-1}). \tag{1.62}$$

Note that we can identify

$$\hat{\mathcal{H}} = \oplus \int_0^\infty \hat{\mathcal{H}}(\lambda)\, d\lambda, \tag{1.63}$$

and that (1.61) is equivalent to

$$F\ A_0\ F^{-1} = \lambda, \tag{1.64}$$

where λ denotes the operator of multiplication by the independent variable in the direct integral (1.63).

Remark that it follows from (1.64) that the spectrum of A_0, $\sigma(A_0)$ is given by

$$\sigma(A_0) = [0, \infty), \tag{1.65}$$

and that A_0 is absolutely continuous.

§2. The Limiting Absorption Principle for the Unpertubed Acoustic Propagator

In this section we will prove the limiting absorption principle for the unperturbed acoustic propagator.

The basic input in the proof is the existence and the Hölder continuity of the trace maps, stated below.

The existence and the continuity of these maps tells us that we can take a sharp frequency (the spectral parameter) in a Hölder continuous way. This allows us to characterize in a proper way the fact that the spectral family of A_0 has particular regularity properties. These regularity properties express in a deep and subtle way the physical content of the theory described by A_0. In fact the physics of our problem is built into the existence and the Hölder continuity of the trace maps.

We first introduce the appropriate spaces of functions.

We denote by $L_s^2(\mathbf{R}^{n+1})$, $s \in \mathbf{R}$, the weighted space consisting of all complex valued measurable functions, $f(x, y)$, on $\mathbf{R}^{n+1}$ such that

$$\eta_s(x, y)\ f(x, y) \in L^2(\mathbf{R}^{n+1}), \tag{2.1}$$

with the natural norm

$$\|f\|_{L_s^2(\mathbf{R}^{n+1})} = \|\eta_s\ f\|_{L^2(\mathbf{R}^{n+1})}, \tag{2.2}$$

where

$$\eta_s(x, y) = (1 + |x|^2 + y^2)^{s/2}. \tag{2.3}$$

By $C_0^\infty(\mathbf{R}^{n+1})$ we denote the space of all complex valued infinitely differentiable functions on $\mathbf{R}^{n+1}$ with compact support.

Lemma 2.1

For each $\lambda > 0$ and $s > \frac{1}{2}$ there is a trace map, $T_0(\lambda)$, bounded from $L_s^2(\mathbf{R}^{n+1})$ into $L^2(S_c)$ such that for every $\varphi \in C_0^\infty(\mathbf{R}^{n+1})$,

$$(T_0(\lambda)\varphi)(\omega) = \int \overline{\phi}_0(x, y, \lambda, \omega)\ \varphi(x, y)\ c_0^{-2}(y)\ dxdy. \tag{2.4}$$

The function $\lambda \to T_0(\lambda)$ from $(0, \infty)$ into $\mathcal{B}(L_s^2(\mathbf{R}^{n+1}),\ L^2(S_c))$ is locally Hölder continuous with exponent γ, where $\gamma < 1$, $\gamma < (s - 1/2)$, if $\lambda \neq \lambda_j$, $j = 1, 2, 3, \cdots$, and $\gamma < 1/2$, $\gamma < (s - 1/2)$, if $\lambda = \lambda_j$, for some $j = 1, 2, 3, \cdots$.

Proof: See Appendix 1, Lemma 1.1.

Q. E. D.

Lemma 2.2

For each $j = 1, 2, 3, \cdots$, $\lambda \in (\lambda_j, \infty)$ and $s > \frac{1}{2}$ there is a trace map, $T_j(\lambda)$, bounded from $L^2_s(\mathbf{R}^{n+1})$ into $L^2(S_1^{n-1})$ such that for every $\varphi \in C_0^\infty(\mathbf{R}^{n+1})$, $\lambda > \lambda_j$,

$$(T_j(\lambda)\varphi)(\nu) = \int \overline{\phi}_j(x, y, \lambda, \nu)\ \varphi(x, y)\ c_0^{-2}(y)\ dxdy. \tag{2.5}$$

Moreover the function $\lambda \to T_j(\lambda)$ from (λ_j, ∞) into $\mathcal{B}(L^2_s(\mathbf{R}^{n+1}), L^2(S_1^{n-1}))$ is locally Hölder continuous with exponent $\gamma < 1$, $\gamma < (s - 1/2)$. Furthermore if we extend $T_j(\lambda)$ to $\lambda = \lambda_j$ as $T_j(\lambda_j) = 0$, for $j = 2, 3, 4, \cdots$, and if $c_+ < c_-$ $T_1(\lambda_1) = 0$, then $T_j(\lambda)$ is Hölder continuous at $\lambda = \lambda_j$ with exponent $\gamma \leq 1/2$, $\gamma < (s - 1/2)$.

Proof: See Appendix 1, Lemma 1.2.

Q. E. D.

We denote

$$\hat{\mathcal{H}}(\infty) = L^2(S_c) \bigoplus_{i=1}^{\infty} L^2(S_1^{n-1}). \tag{2.6}$$

Note that $\hat{\mathcal{H}}(\lambda)$ is naturally imbedded into $\hat{\mathcal{H}}(\infty)$ if we identify $\varphi \in \hat{\mathcal{H}}(\lambda)$ with the vector in $\hat{\mathcal{H}}(\infty)$ that has the same components in $L^2(S_c)$ and in the first j copies of $L^2(S_1^{n-1})$ as φ, where $\lambda_j < \lambda \leq \lambda_{j+1}$, and all the others are zero. In this sense $\hat{\mathcal{H}}(\lambda) \subset \hat{\mathcal{H}}(\infty)$, for $\lambda \in (0, \infty)$.

For $\lambda \in (0, \infty)$, we denote by $B(\lambda)$ the following operator from $\mathcal{H}_0$ into $\hat{\mathcal{H}}(\infty)$

$$B(\lambda)\varphi = \bigoplus_{j=0}^{\infty} T_j(\lambda)\ \eta_{-s}\varphi, \tag{2.7}$$

where $s > 1/2$, and where we extend the definition of $T_j(\lambda)$, $j = 1, 2, 3, \cdots$, to $\lambda \in (0, \lambda_j]$ by zero, i.e.

$$T_j(\lambda) = 0,\ \lambda \in (0, \lambda_j], \tag{2.8}$$

$j = 1, 2, 3, \cdots$. Note that since $T_j(\lambda_j) = 0$, the operator valued function $T_j(\lambda)$ remains locally Hölder continuous at $\lambda = \lambda_j$, with the same exponent as in Lemma 2.2, and it is obviously locally Hölder continuous with exponent equal to one for $\lambda \in (0, \lambda_j)$.

Lemma 2.3

The operator $B(\lambda)$ defined in (2.7) is bounded from $\mathcal{H}_0$ into $\hat{\mathcal{H}}(\infty)$, and the operator valued function $\lambda \to B(\lambda)$ from $(0, \infty)$ into $\mathcal{B}(\mathcal{H}_0, \hat{\mathcal{H}}(\infty))$ is locally Hölder continuous with exponent γ satisfying $\gamma < 1$, $\gamma < (s - 1/2)$, if $\lambda \neq \lambda_j$, $j = 1, 2, 3, \cdots$, and $\gamma < 1/2$, $\gamma < (s - 1/2)$ if $\lambda = \lambda_j$, for some $j = 1, 2, 3, \cdots$.

Proof: For any $\lambda \in (0, \infty)$ there is a j such that $\lambda_j < \lambda \leq \lambda_{j+1}$. It follows that all the terms in (2.7) with $i \geq j + 1$ are zero and

$$B(\lambda)\varphi = \bigoplus_{i=0}^{j} T_i(\lambda)\, \eta_{-s}\, \varphi \in \hat{\mathcal{H}}(\lambda). \tag{2.9}$$

The lemma follows now from Lemmas 2.1 and 2.2 since $\hat{\mathcal{H}}(\lambda) \subset \hat{\mathcal{H}}(\infty)$.

Q. E. D.

Note that it follows from the proof of Lemma 2.3 that $B(\lambda)$ is bounded from $\mathcal{H}_0$ into $\hat{\mathcal{H}}(\lambda)$.

It follows from Theorem 1.1, (1.61), and Lemmas 2.1, 2.2, and 2.3, that for every Borel set $\Delta \subset (0, \infty)$, and $s > 1/2$

$$F\, E_0(\Delta)\, \eta_{-s} = \chi_\Delta(\lambda)\, B(\lambda), \tag{2.10}$$

where $\chi_\Delta(\lambda)$ denotes the characteristic function of Δ, and $E_0(\cdot)$ is the family of spectral projectors of A_0.

For $z \in \mathbf{C} \setminus [0, \infty)$ we denote by

$$R_0(z) = (A_0 - z)^{-1}, \tag{2.11}$$

the resolvent of A_0.

For any $z = \lambda + i\,\epsilon$, $\lambda \in (0, \infty)$, $\epsilon \neq 0$, let I_λ be any compact interval, $I_\lambda \subset (0, \infty)$, such that λ is an interior point of I_λ. It follows from (2.10) that

$$\eta_{-s} R_0(z) \eta_{-s} = \int_{I_\lambda} \frac{1}{\rho - z}\, B^*(\rho) B(\rho)\, d\rho + \eta_{-s} E_0(I_\lambda^{\sim}) R_0(z) \eta_{-s}, \tag{2.12}$$

where $I_\lambda^\sim$ denotes the complement of I_λ.

For m any nonnegative integer and s any real number we denote by $H_{m,s}(\mathbf{R}^{n+1})$ the weighted Sobolev space consisting of all complex valued measurable functions, $\varphi(x,y)$, on $\mathbf{R}^{n+1}$ such that

$$\eta_s(x,y)\ \varphi(x,y) \in H_m(\mathbf{R}^{n+1}), \tag{2.13}$$

with the natural norm

$$\|\varphi\|_{H_{m,s}(\mathbf{R}^{n+1})} = \|\eta_s\varphi\|_{H_m(\mathbf{R}^{n+1})}, \tag{2.14}$$

where $\eta_s(x,y)$ is as in (2.3).

Note that if $s_1 \geq s_2$, then for any m, $H_{m,s_1}(\mathbf{R}^{n+1}) \subset H_{m,s_2}(\mathbf{R}^{n+1})$, with a continuous imbedding. Furthermore $H_{m,0}(\mathbf{R}^{n+1}) = H_m(\mathbf{R}^{n+1})$.

For any $s \geq 0$ the norm of $H_{2,-s}(\mathbf{R}^{n+1})$ is equivalent to the norm

$$\|(I-\Delta)\eta_{-s}\varphi\|_{L^2(\mathbf{R}^{n+1})}, \tag{2.15}$$

for $\varphi \in H_{2,-s}(\mathbf{R}^{n+1})$. Moreover, since for $z \in \mathbf{C} \setminus [0,\infty)$

$$A_0 R_0(z) = I + z\ R_0(z), \tag{2.16}$$

$$\|R_0(z)\varphi\|_{H_2(\mathbf{R}^{n+1})} \leq C\|(I-\Delta)R_0(z)\varphi\|_{L^2(\mathbf{R}^{n+1})} \leq$$

$$\leq C(\|\varphi\|_{L^2(\mathbf{R}^{n+1})} + \|R_0(z)\varphi\|_{L^2(\mathbf{R}^{n+1})}), \tag{2.17}$$

and it follows that $R_0(z)$ is an analytic function of $z \in \mathbf{C} \setminus [0,\infty)$ with values on $\mathcal{B}(L^2(\mathbf{R}^{n+1}),\ \ H_2(\mathbf{R}^{n+1}))$. Then also $R_0(z)$ is an analytic function of $z \in \mathbf{C} \setminus [0,\infty)$ with values on $\mathcal{B}(L^2_s(\mathbf{R}^{n+1}),\ H_{2,-s}(\mathbf{R}^{n+1}))$, for any $s \geq 0$.

We denote

$$\mathbf{C}^{\pm} = \{z \in \mathbf{C}\ :\ \pm\, Im\ z > 0\}. \tag{2.18}$$

We now state our result on the limiting absorption principle for A_0.

Theorem 2.4

A_0 is an absolutely continuous operator with spectrum, $\sigma(A_0) = [0,\infty)$. Let us take $s > 1/2$. Then for every $\lambda > 0$ the following limits

$$R_0(\lambda \pm i\ 0) = \lim_{\epsilon \downarrow 0} R_0(\lambda \pm i\ \epsilon), \tag{2.19}$$

exist in the uniform operator topology on $\mathcal{B}(L^2_s(\mathbf{R}^{n+1}),\ H_{2,-s}(\mathbf{R}^{n+1}))$. Moreover the convergence is uniform for λ in compact sets of $(0,\infty)$.

The functions

$$R_0^{\pm}(\lambda) = \begin{cases} R_0(\lambda), & Im\ \lambda \neq 0, \\ R_0(\lambda \pm i\,0), & Im\ \lambda = 0, \end{cases} \tag{2.20}$$

defined for $\lambda \in \mathbf{C}^{\pm} \cup (0,\infty)$ with values on $\mathcal{B}(L_s^2(\mathbf{R}^{n+1}),\ H_{2,-s}(\mathbf{R}^{n+1}))$, are analytic for $Im\ \lambda \neq 0$, and locally Hölder continuous for $Im\ \lambda = 0$, with exponent γ satisfying $\gamma < 1,\ \gamma < (s - 1/2)$, if $\lambda \neq \lambda_j,\ j = 1,2,3,\cdots$, and $\gamma < 1/2,\ \gamma < (s-1/2)$, if $\lambda = \lambda_j$, for some $j = 1,2,3,\cdots$.

Moreover the $R_0(\lambda \pm i\,0)$ are given by

$$\eta_{-s} R_0(\lambda \pm i\,0)\,\eta_{-s} = P.V. \int_{I_\lambda} \frac{1}{\rho - \lambda} B^*(\rho)B(\rho)d\rho \pm i\,\pi\, B^*(\lambda)B(\lambda)$$

$$+\,\eta_{-s} E_0(I_\lambda^{\sim}) R_0(\lambda)\eta_{-s}, \tag{2.21}$$

where $P.V.$ stands for the principal value of the integral.

Proof: If we consider the uniform operator topology $\mathcal{B}(L_s^2(\mathbf{R}^{n+1}),\ L_{-s}^2(\mathbf{R}^{n+1}))$ the theorem follows (with the exception of the absolute continuity of A_0) from Lemma 2.3 and a classical theorem in the existence of boundary values in the integral in (2.12) (see Muskhelishvili 1953, Privalov 1956, and Kuroda 1980, Theorem 5 and Remark 6 in section 4.1).

Notice that by functional calculus the existence and the Hölder continuity of the limits when ϵ goes to zero is trivial for the second term in (2.12).

To prove the theorem in the uniform operator topology in $\mathcal{B}(L_s^2(\mathbf{R}^{n+1}),\ H_{2,-s}(\mathbf{R}^{n+1}))$ we recall that the norm of $H_{2,-s}(\mathbf{R}^{n+1})$ is equivalent to the norm in (2.15). This last norm is in turn equivalent to the norm

$$\|\varphi\|_{L_{-s}^2(\mathbf{R}^{n+1})} + \|A_0\varphi\|_{L_{-s}^2(\mathbf{R}^{n+1})}. \tag{2.22}$$

But by (2.16)

$$\|R_0(\lambda \pm i\,\epsilon)\varphi\|_{H_{2,-s}(\mathbf{R}^{n+1})} \leq C\,[\,\|R_0(\lambda \pm i\,\epsilon)\varphi\|_{L_{-s}^2(\mathbf{R}^{n+1})} +$$

$$+\,|\lambda \pm i\,\epsilon|\,\|R_0(\lambda \pm i\,\epsilon)\varphi\|_{L_{-s}^2(\mathbf{R}^{n+1})}], \tag{2.23}$$

and the results in $\mathcal{B}(L_s^2(\mathbf{R}^{n+1}),\ H_{2,-s}(\mathbf{R}^{n+1}))$ follow from the ones already proven in $\mathcal{B}(L_s^2(\mathbf{R}^{n+1}),\ L_{-s}^2(\mathbf{R}^{n+1}))$.

The absolute continuity of A_0 follows from (1.64) or from the existence of the limits in (2.19) uniformly on compact sets of $(0,\infty)$ (see Reed and Simon 1978, Section XIII.6).

The fact that $\sigma(A_0) = [0, \infty)$ follows from (1.64).

Q. E. D.

§3. The Perturbed Acoustic Propagator

In this section we study the acoustic propagator of a perturbed stratified fluid.

The simplest form of perturbation is when the fluid occupies all space, but locally is not stratified. The wave equation in this case is

$$\frac{\partial^2}{\partial_t^2} u(x, y, t) - c^2(x, y)\Delta\, u(x, y, t) = 0, \tag{3.1}$$

where $c(x, y)$ is a real valued measurable function on $\mathbf{R}^{n+1}$ that is bounded below and above

$$0 < c_m \leq c(x, y) \leq c_M, \tag{3.2}$$

for some positive constants c_m and c_M.

The function $c(x, y)$ is asymptotic to $c_0(y)$ when $|x| + |y| \to \infty$ in a sense that we will specify in later sections.

We denote by $\mathcal{H}$ the Hilbert space consisting of all functions on $L^2(\mathbf{R}^{n+1})$ with the scalar product

$$(\varphi, \psi)_{\mathcal{H}} = \int_{\mathbf{R}^{n+1}} \varphi(x, y)\overline{\psi}(x, y)\; c^{-2}(x, y)\; dxdy. \tag{3.3}$$

Note that because of (3.2) the norms of $\mathcal{H}$ and of $L^2(\mathbf{R}^{n+1})$ are equivalent.

The perturbed acoustic propagator in all space is

$$A = -c^2(x, y)\Delta, \tag{3.4}$$

with domain

$$D(A) = H_2(\mathbf{R}^{n+1}). \tag{3.5}$$

Since $-\Delta$ is selfadjoint and positive in $L^2(\mathbf{R}^{n+1})$, A is selfadjoint and positive in $\mathcal{H}$.

We also consider the situation when the fluid is in the presence of an obstacle. In this case the problem is defined in an exterior domain, that is to say in an open set $\Omega_e \subset \mathbf{R}^{n+1}$ such that $\mathbf{R}^{n+1} \setminus \Omega_e$ is compact and its topological boundary, $\partial\Omega_e$, is of measure zero on $\mathbf{R}^{n+1}$. The appropriate wave equation is

$$\frac{\partial^2}{\partial_t^2} u(x, y, t) - c^2(x, y)\Delta\, u(x, y, t) = 0, \tag{3.6}$$

where now the acoustic potential, $u(x, y, t)$, is only defined in Ω_e. The speed $c(x, y)$ is also defined only on Ω_e and is real valued and measurable, and satisfies (3.2) on Ω_e.

For Ω any open set, $\Omega \subset \mathbf{R}^{n+1}$, we denote by $L^2(\Omega)$ the Hilbert space of all the complex valued Lebesgue square integrable functions on Ω, with scalar product

$$(\varphi, \psi)_{L^2(\Omega)} = \int_\Omega \varphi(x, y)\overline{\psi}(x, y)\, dxdy. \tag{3.7}$$

By $C_0^\infty(\Omega)$ we denote the set of all complex valued infinitely differentiable functions with compact support in Ω.

The operator Δ is symmetric and negative when defined on $C_0^\infty(\Omega_e)$. We denote by Δ_e any selfadjoint, bounded above operator in $L^2(\Omega_e)$, that is an extension of Δ in $C_0^\infty(\Omega_e)$.

We denote by $\mathcal{H}_e$ the Hilbert space consisting of all functions in $L^2(\Omega_e)$ with scalar product

$$(\varphi, \psi)_{\mathcal{H}_e} = \int_{\Omega_e} \varphi(x, y)\overline{\psi}(x, y)\; c^{-2}(x, y)\, dxdy. \tag{3.8}$$

The perturbed acoustic propagator in the exterior domain is

$$A_e = -c^2(x, y)\Delta_e, \tag{3.9}$$

with domain

$$D(A_e) = D(\Delta_e). \tag{3.10}$$

A_e is selfadjoint and bounded below on $\mathcal{H}_e$. Note that the norms of $L^2(\Omega_e)$ and of $\mathcal{H}_e$ are equivalent.

We will assume that Δ_e satisfies the following local compactness assumption.

Assumption 3.1

We say that Δ_e satisfies the local compactness assumption if for some $f(x, y) \in C_0^\infty(\mathbf{R}^{n+1})$, with $f \equiv 1$ in a neighborhood of $\partial\Omega_e$, and some $M > 0$ such that $-\Delta_e + M > 0$, the operator

$$f(x, y)(-\Delta_e + M)^{-1}, \tag{3.11}$$

is compact on $L^2(\Omega_e)$.

Note that it follows from the first resolvent equation (see Kato 1976, Chapter 5, problem 5.4, and page 173) that

$$(-\Delta_e - z_1)^{-1} - (-\Delta_e - z_2)^{-1} = (z_2 - z_1)(-\Delta_e - z_1)^{-1}(-\Delta_e - z_2)^{-1}, \tag{3.12}$$

for $z_1, z_2 \in \mathbf{C} \setminus \sigma(-\Delta_e)$, that if Assumption 3.1 holds, then

$$f(x,y)(-\Delta_e - z)^{-1}, \tag{3.13}$$

is compact on $L^2(\Omega_e)$ for any $z \in \rho(-\Delta_e)$.

Moreover let $f'(x,y)$ be any function in $C_0^\infty(\mathbf{R}^{n+1})$ such that $f'(x,y) \equiv 1$ in a neighborhood of $\partial\Omega_e$. Then $(f(x,y) - f'(x,y)) \in C_0^\infty(\mathbf{R}^{n+1})$, $f(x,y) - f'(x,y) \equiv 0$ in a neighborhood of $\partial\Omega_e$. It follows from Lemma 4.3 and Remark 4.7 in section 4 that (note that by Rellich local compactness theorem (see Adams 1975) the imbedding of $H_{2,s_1}(\mathbf{R}^{n+1})$ into $L^2_{s_2}(\mathbf{R}^{n+1})$ is compact if $s_1 > s_2 \geq 0$),

$$(f(x,y) - f'(x,y))(-\Delta_e + M)^{-1}, \tag{3.14}$$

is compact in $L^2(\Omega_e)$.

Then if Assumption 3.1 is satisfied, the operator

$$f(x,y)(-\Delta_e - z)^{-1}, \tag{3.15}$$

is compact in $L^2(\Omega_e)$ for any function $f(x,y) \in C_0^\infty(\mathbf{R}^{n+1})$, with $f(x,y) \equiv 1$ in a neighborhood of $\partial\Omega_e$ and any $z \in \rho(-\Delta_e)$.

Finally we consider the transmission problem. Let Ω_i be a bounded open set in $\mathbf{R}^{n+1}$, whose topological boundary $\partial\Omega_i$ is of measure zero in $\mathbf{R}^{n+1}$. Denote $\Omega_e = \mathbf{R}^{n+1} \setminus \overline{\Omega}_i$, where $\overline{\Omega}_i$ denotes the closure of Ω_i, and

$$\Omega = \Omega_i \cup \Omega_e \equiv \mathbf{R}^{n+1} \setminus \partial\Omega_i. \tag{3.16}$$

The appropriate wave equation is

$$\frac{\partial^2}{\partial_t^2} u(x,y,t) - c^2(x,y)\, \Delta u(x,y,t) = 0, \tag{3.17}$$

now in distribution sense in Ω. $c(x,y)$ is real valued, measurable, and satisfies (3.2) in $\mathbf{R}^{n+1}$.

We denote by Δ_T any operator in $L^2(\mathbf{R}^{n+1})$ that is a selfadjoint bounded above extension of Δ in $C_0^\infty(\Omega)$.

The perturbed acoustic propagator for the transmission problem is

$$A_T = -c^2(x,y)\Delta_T, \tag{3.18}$$

with domain

$$D(A_T) = D(\Delta_T). \tag{3.19}$$

A_T is selfadjoint and bounded below in $\mathcal{H}$. Note that since $\mathbf{R}^{n+1} \setminus \Omega$ is of measure zero, $L^2(\mathbf{R}^{n+1}) \equiv L^2(\Omega)$.

We will assume that Δ_T satisfies the following local compactness assumption.

Assumption 3.2

We say that Δ_T satisfies the local compactness assumption if for some $f(x,y) \in C_0^\infty(\mathbf{R}^{n+1})$, with $f(x,y) \equiv 1$ in a neighborhood of $\partial\Omega$, and some $M > 0$ such that $-\Delta_T + M > 0$, the operator

$$f(x,y)(-\Delta_T + M)^{-1}, \tag{3.20}$$

is compact on $L^2(\mathbf{R}^{n+1})$.

As in the case of exterior domains we prove that if Assumption 3.2 is satisfied, then the operator

$$f(x,y)(-\Delta_T - z)^{-1} \tag{3.21}$$

is compact for any $f(x,y) \in C_0^\infty(\mathbf{R}^{n+1})$, with $f(x,y) \equiv 1$ in a neighborhood of $\partial\Omega$, and any $z \in \rho(-\Delta_T)$.

In practice, the selfadjoint bounded above extensions Δ_e and Δ_T are constructed by imposing appropriate boundary conditions respectively on $\partial\Omega_e$ and $\partial\Omega$.

The local compactness assumption is a restriction on the regularity of the boundary and on the boundary condition.

It implies the decay in time of the local energy and it is satisfied by most of the boundaries and boundary conditions that are encountered in the applications.

We give below several examples.

For Ω any open set in $\mathbf{R}^{n+1}$, and m any non-negative integer, we denote by $H_m(\Omega)$ the Sobolev space of order m consisting of all the complex valued functions that are Lebesgue square integrable on Ω, all of whose distributional derivatives of order up to m are also square integrable, with scalar product

$$(\varphi, \psi)_{H_m(\Omega)} = \sum_{|\alpha| \leq m} (D^\alpha \varphi, \quad D^\alpha \psi)_{L^2(\Omega)}. \tag{3.22}$$

We denote by $H_m^0(\Omega)$ the completion of $C_0^\infty(\Omega)$ in the norm of $H_m(\Omega)$.

Example 3.3 (Dirichlet Boundary Condition)

We denote by $h_{e,D}$ the following quadratic form

$$h_{e,D}(\varphi, \psi) = \sum_{i=1}^{n} \left(\frac{\partial}{\partial x_i} \varphi, \frac{\partial}{\partial x_i} \psi \right)_{L^2(\Omega_e)} + \left(\frac{\partial}{\partial y} \varphi, \frac{\partial}{\partial y} \psi \right)_{L^2(\Omega_e)}, \tag{3.23}$$

with domain

$$D(h_{e,D}) = H_1^0(\Omega_e) \times H_1^0(\Omega_e). \tag{3.24}$$

The form $h_{e,D}$ is closed symmetric and nonnegative. Let $-\Delta_{e,D}$ be the associated selfadjoint nonnegative operator (see Kato 1976, Chapter 6). $\Delta_{e,D}$ is the selfadjoint extension of Δ with Dirichlet boundary condition, since, as $D(\Delta_{e,D}) \subset H_1^0(\Omega_e)$, the functions on $D(\Delta_{e,D})$ are zero on $\partial\Omega_e$ in trace sense.

Moreover, since the form domain of $\Delta_{e,D}$ is $H_1^0(\Omega_e)$, $(-\Delta_{e,D} + M)^{-1}$ is a bounded operator from $L^2(\Omega_e)$ into $H_1^0(\Omega_e)$, for any $M > 0$. Then Assumption 3.1 is satisfied since by the Rellich local compactness theorem (see Adams 1975) the imbedding of $H_1^0(\Omega_e)$ into $L^2(\Omega_e \cap B_N)$ is compact for any $N > 0$, where by B_N we denote the ball of radius N in $\mathbf{R}^{n+1}$. Note that in this case no regularity of the boundary is required.

In the case of the transmission problem the extension $-\Delta_{T,D}$, with Dirichlet boundary condition is defined as the selfadjoint non-negative operator in $L^2(\mathbf{R}^{n+1})$ associated with the form

$$h_{T,D}(\varphi, \psi) = \sum_{i=1}^{n} \left(\frac{\partial}{\partial x_i} \varphi, \frac{\partial}{\partial x_i} \psi \right)_{L^2(\Omega)} + \left(\frac{\partial}{\partial y} \varphi, \frac{\partial}{\partial y} \psi \right)_{L^2(\Omega)}, \tag{3.25}$$

with domain

$$D(h_{T,D}) = H_1^0(\Omega) \times H_1^0(\Omega). \tag{3.26}$$

We prove that Assumption 3.2 is satisfied as in the case of exterior domains.

Example 3.4 (Generalized Neumann Boundary Condition)

We consider now the form (3.23), but with domain $H_1(\Omega_e) \times H_1(\Omega_e)$. The associated selfadjoint nonnegative operator, $\Delta_{e,N}$, is the selfadjoint extension of $-\Delta$ in $C_0^\infty(\Omega)$ with generalized Neumann boundary condition. If $\partial\Omega_e$ is regular enough the functions, φ, in the domain of $\Delta_{e,N}$ satisfy the classical Neumann condition in trace sense

$$\frac{\partial \varphi}{\partial \nu} = 0, \tag{3.27}$$

where ν denotes the normal vector to $\partial\Omega$. Since the form domain of $\Delta_{e,N}$ is $H_1(\Omega_e)$, Assumption 3.1 will hold if the imbedding of $H_1(\Omega_e)$ into $L^2(\Omega_e \cap B_N)$ is compact for any $N > 0$. Sufficient conditions for this to be true are known. See Adams 1975, and Wilcox 1975.

Example 3.5 (Generalized Robin Boundary Condition)

Suppose that $\partial\Omega_e$ is regular enough in order that there is a bounded trace operator from $H_1(\Omega_e)$ into $L^2(\partial\Omega_e)$ (see Lions and Magenes 1972, and Reed and Simon 1975), that is to say a bounded operator, T, from $H_1(\Omega_e)$ into $L^2(\partial\Omega_e)$, such that

$$(T\varphi)(x,y) = \varphi(x,y), \tag{3.28}$$

for all $\varphi(x,y) \in C^\infty(\Omega_e)$.

Let $\sigma(x,y)$ be a bounded measurable function on $\partial\Omega_e$. We define the quadratic form

$$h_\sigma(\varphi,\, \psi) = \sum_{i=1}^{n} \left(\frac{\partial}{\partial x_i}\varphi,\, \frac{\partial}{\partial x_i}\psi \right)_{L^2(\Omega_e)} + \left(\frac{\partial}{\partial y}\,\varphi,\, \frac{\partial}{\partial y}\,\psi \right)_{L^2(\Omega_e)}$$

$$+ \int_{\partial\Omega_e} \sigma(x,y)(T\varphi)(x,y)\overline{(T\psi(x,y))}d\omega, \tag{3.29}$$

where $d\omega$ denotes the measure induced on $\partial\Omega_e$ by Lebesgue measure on $\mathbf{R}^{n+1}$.

As it is easily checked h_σ is closed, symmetric and bounded below. Let $-\Delta_{e,\sigma}$ be the associated selfadjoint, bounded below operator. The local compactness assumption is satisfied under the same conditions as in Example 3.4. Note that the form domain is $H_1(\Omega_e)$ in both cases.

If $\partial\Omega_e$ is regular enough the functions in the domain of $\Delta_{e,\sigma}$ satisfy the classical Robin boundary condition

$$\frac{\partial\varphi}{\partial\nu} = \sigma\,\varphi, \tag{3.30}$$

in $\partial\Omega_e$ in trace sense. By ν we denote the normal vector pointing towards Ω_e.

Example 3.6 (Transmission Condition)

We consider the form (3.29) with domain $H_1(\mathbf{R}^{n+1})$. Let $-\Delta_{T,\sigma}$ be the associated selfadjoint bounded below operator. Since the form domain is $H_1(\mathbf{R}^{n+1})$ the local compactness assumption is satisfied by the Rellich local compactness theorem (see Adams 1975, and Wilcox 1975).

If $\partial\Omega$ is regular enough the functions on the domain of $\Delta_{T,\sigma}$ satisfy the following transmission condition on $\partial\Omega$ in trace sense

$$\left[\frac{\partial\varphi}{\partial\nu}\right] = \sigma\,\varphi, \tag{3.31}$$

where $\left[\frac{\partial\varphi}{\partial\nu}\right]$ denotes the jump in the normal derivative accros $\partial\Omega$.

Finally note that the perturbed propagator in all space (3.4) is the particular case where $\overline{\Omega}_i$ is the empty set. The local compactness Assumption 3.2 holds by the Rellich local compactness theorem in all space.

For a further discussion on boundary conditions and related problems, see Combes and Weder 1981, and Weder 1984.

§4. The Essential Spectrum of the Perturbed Acoustic Propagator

In this section we prove that the difference of the resolvents of the perturbed and unperturbed acoustic propagator is a compact operator between appropriate weighted spaces. This result will be used in the proof of the limiting absorption principle in Section 6.

As a consequence of the compactness of the difference of resolvents we prove that the essential spectrum of the perturbed acoustic propagator consists of $[0,\infty)$.

We first state some preliminary results.

Lemma 4.1

For any $s \geq 0$, $(A_0 + M)^{-1}$, $M > 0$, is a bounded operator from $L^2_s(\mathbf{R}^{n+1})$ into $H_{2,s}(\mathbf{R}^{n+1})$.

Proof: It is enough to prove that for any multiindex α, with $|\alpha| \leq 2$, the operator

$$\eta_s(x,y)\ D^\alpha (A_0+M)^{-1}\ \eta_{-s}(x,y) \tag{4.1}$$

is a bounded operator on $L^2(\mathbf{R}^{n+1})$. But this is equivalent to proving that

$$\eta_{s,N}(x,y)\ D^\alpha (A_0+M)^{-1}\ \eta_{-s,N}(x,y) \tag{4.2}$$

is bounded on $L^2(\mathbf{R}^{n+1})$, for $N > 0$, where

$$\eta_{s,N}(x,y) = \left(1 + \frac{|x|^2}{N^2} + \frac{|y|^2}{N^2}\right)^{s/2}. \tag{4.3}$$

We have that

$$(A_0+M)^{-1}\eta_{-s,N} = \eta_{-s,N}(A_0+M)^{-1} + (A_0+M)^{-1}\eta_{-s,N}\eta_{s,N}$$

$$\cdot\ [\eta_{-s,N}, A_0]\ (A_0+M)^{-1}. \tag{4.4}$$

Moreover

$$\|\eta_{s,N}[\eta_{-s,N}, A_0]\ (A_0+M)^{-1}\|_{\mathcal{B}(L^2(\mathbf{R}^{n+1}))} \leq \frac{C}{N}, \tag{4.5}$$

for some constant C. Then for N large enough

$$\begin{aligned}(A_0+M)^{-1}\eta_{-s,N} &= \eta_{-s,N}(A_0+M)^{-1}\\ &\cdot (I-\eta_{s,N}[\eta_{-s,N}, A_0]\ (A_0+M)^{-1})^{-1}.\end{aligned} \tag{4.6}$$

It follows that

$$\begin{aligned}\eta_{s,N}D^{\alpha}(A_0+M)^{-1}\eta_{-s,N} &= \eta_{s,N}D^{\alpha}\eta_{-s,N}(A_0+M)^{-1}\\ &\cdot (I-\eta_{s,N}[\eta_{-s,N}, A_0]\ (A_0+M)^{-1})^{-1}.\end{aligned} \tag{4.7}$$

The lemma follows since the right hand side of (4.7) is a bounded operator on $L^2(\mathbf{R}^{n+1})$.

Q. E. D.

Lemma 4.2

Suppose that some $\varphi \in \mathcal{H}_0$ satisfies

$$(\varphi,\ (A_0+M)\psi)_{\mathcal{H}_0} = 0, \tag{4.8}$$

for every $\psi \in C_0^{\infty}(\mathbf{R}^{n+1} \setminus \partial\Omega)$ and some $M > 0$. Then for every $f \in C_0^{\infty}(\mathbf{R}^{n+1})$, $f \equiv 1$ in a neighborhood of $\partial\Omega$, $(1-f)\,\varphi \in H_{2,s}(\mathbf{R}^{n+1})$, for all $s \geq 0$, and

$$\|(1-f)\varphi\|_{H_{2,s}(\mathbf{R}^{n+1})} \leq C\ \|\varphi\|_{\mathcal{H}_0}, \tag{4.9}$$

where the constant C depends only on f and s.

Proof: Equation (4.8) implies that

$$-\Delta\varphi = -M\ c_0^{-2}(y)\varphi, \tag{4.10}$$

in distribution sense in Ω. It follows by interior elliptic regularity that $(1-f)\varphi \in H_2(\mathbf{R}^{n+1})$. Moreover by (4.10)

$$(A_0+M)(1-f)\varphi = (-A_0 f)\varphi + 2\ c_0^2(y)\nabla f\cdot\nabla\varphi, \tag{4.11}$$

where $\nabla = \left(\frac{\partial}{\partial x_1}, \frac{\partial}{\partial x_2}, \cdots, \frac{\partial}{\partial x_n}, \frac{\partial}{\partial y} \right)$ is the gradient operator. By interior elliptic regularity, and since $(A_0 f)$ and ∇f have compact support on $\mathbf{R}^{n+1} \setminus \partial\Omega$, it follows that the right hand side of (4.11) belongs to $L^2_s(\mathbf{R}^{n+1})$, for all $s \geq 0$. Then by Lemma 4.1 $(1-f)\varphi \in H_{2,s}(\mathbf{R}^{n+1})$, and

$$\|(1-f)\varphi\|_{H_{2,s}(\mathbf{R}^{n+1})} \leq C\, \|\varphi\|_{\mathcal{H}_0}, \tag{4.12}$$

where we used the fact that by interior elliptic regularity

$$\|\nabla f \cdot \nabla\varphi\|_{L^2_s(\mathbf{R}^{n+1})} \leq C\, \|\varphi\|_{\mathcal{H}_0}. \tag{4.13}$$

Q. E. D.

We denote by $A_{T,0}$ the following selfadjoint bounded below operator on $\mathcal{H}_0$

$$A_{T,0} = -c_0^2(y)\Delta_T, \tag{4.14}$$

with domain

$$D(A_{T,0}) = D(\Delta_T). \tag{4.15}$$

We have that

<u>Lemma 4.3</u>

Let $M > 0$ be such that $A_{T,0} + M > 0$. Then for any $f(x,y) \in C_0^\infty(\mathbf{R}^{n+1})$, such that $f \equiv 1$ in a neighborhood of $\partial\Omega$, the operator $(1-f)(A_{T,0}+M)^{-1}$ is bounded from $L^2_s(\mathbf{R}^{n+1})$ into $H_{2,s}(\mathbf{R}^{n+1})$, for any $s \geq 0$.

<u>Proof</u>: Let $\psi \in L^2_s(\mathbf{R}^{n+1})$, and denote

$$\varphi = (A_{T,0} + M)^{-1}\psi. \tag{4.16}$$

Then in distribution sense in Ω

$$\Delta\varphi = c_0^{-2}(M\varphi - \psi). \tag{4.17}$$

It follows by interior elliptic regularity that $(1-f)\varphi \in H_2(\mathbf{R}^{n+1})$. Then by (4.16)

$$(A_0 + M)(1-f)\varphi = (-A_0 f)\varphi + 2\, c_0^2(y)\nabla f \cdot \nabla\varphi + (1-f)\psi, \tag{4.18}$$

and by Lemma 4.1

$$\|(1-f)\varphi\|_{H_{2,s}(\mathbf{R}^{n+1})} \leq C \, \|\psi\|_{L^2_s(\mathbf{R}^{n+1})}, \tag{4.19}$$

where we used the fact that $A_0 f$ and ∇f have compact support in Ω, and interior elliptic regularity.

Q. E. D.

We denote by $A_{D,0}$ the following selfadjoint bounded below operator in $\mathcal{H}_0$

$$A_{D,0} = -c_0^2 \Delta_{T,D}, \tag{4.20}$$

with domain

$$D(A_{D,0}) = D(\Delta_{T,D}), \tag{4.21}$$

where $\Delta_{T,D}$ is the transmission Laplacian with Dirichlet boundary condition defined in Section 3.

We have

Lemma 4.4

Let $M > 0$ be such that $A_{T,0} + M > 0$, and suppose that Assumption 3.2 is valid. Then for any s_1, $s_2 \geq 0$, the operator

$$(A_{T,0} + M)^{-1} - (A_{D,0} + M)^{-1}, \tag{4.22}$$

extends to a compact operator from $L^2_{-s_1}(\mathbf{R}^{n+1})$ into $L^2_{s_2}(\mathbf{R}^{n+1})$.

Proof: Let $\chi \in \mathcal{H}_0$ and denote

$$\varphi = ((A_{T,0} + M)^{-1} - (A_{D,0} + M)^{-1})\chi. \tag{4.23}$$

Then since both $A_{T,0}$ and $A_{T,D}$ are extensions of $-c_0^2\Delta$ in $C_0^\infty(\Omega)$ we have that for every $\psi \in C_0^\infty(\Omega)$

$$(\varphi, (A_0 + M)\psi)_{\mathcal{H}_0} = 0. \tag{4.24}$$

By Lemma 4.2 and (4.23) for any $s \geq 0$

$$\|(1-f)((A_{T,0} + M)^{-1} - (A_{D,0} + M)^{-1})\chi\|_{H_{2,s}(\mathbf{R}^{n+1})} \leq C \, \|\chi\|_{\mathcal{H}_0}, \tag{4.25}$$

for every $f(x,y) \in C_0^\infty(\mathbf{R}^{n+1})$, $f(x,y) \equiv 1$ in a neighborhood of $\partial\Omega$. Moreover since the imbedding of $H_{2,s}(\mathbf{R}^{n+1})$ into $L^2_{s_1}(\mathbf{R}^{n+1})$ is compact if $s > s_1$, the operator

$$\eta_{s_1}(x,y)(1-f)((A_{T,0}+M)^{-1} - (A_{D,0}+M)^{-1}), \tag{4.26}$$

is compact on $\mathcal{H}_0$.

By Assumption 3.2 and since $\eta_{s_1}(x,y)$ is bounded in the support of $f(x,y)$,

$$\eta_{s_1}(x,y)f(x,y)(A_{T,0}+M)^{-1}, \tag{4.27}$$

is a compact operator on $\mathcal{H}_0$ (recall that $A_{T,0}$ and Δ_T have the same domain). Since $A_{T,D}$ always satisfies Assumption 3.2 (see Example 3.3), also

$$\eta_{s_1}(x,y)f(x,y)(A_{D,0}+M)^{-1} \tag{4.28}$$

is compact on $\mathcal{H}_0$. By the compactness of the operators in (4.26), (4.27), and (4.28), the operator

$$\eta_{s_1}(x,y)((A_{T,0}+M)^{-1} - (A_{D,0}+M)^{-1}), \tag{4.29}$$

is compact on $\mathcal{H}_0$ for any $s_1 \geq 0$.

By taking adjoints also the operator

$$((A_{T,0}+M)^{-1} - (A_{D,0}+M)^{-1})\eta_{s_1}(x,y), \tag{4.30}$$

is compact on $\mathcal{H}_0$ for all $s_1 \geq 0$. It follows from Hadamard's three lines interpolation theorem (see Reed and Simon 1975, Appendix to IX.4) that for all $s_1,\ s_2 \geq 0$, the operator

$$\eta_{s_2}(x,y)((A_{T,0}+M)^{-1} - (A_{D,0}+M)^{-1})\eta_{s_1}(x,y), \tag{4.31}$$

is compact on $\mathcal{H}_0$. But this is equivalent to claiming that the operator in (4.22) extends to a compact operator from $L^2_{-s_1}(\mathbf{R}^{n+1})$ into $L^2_{s_2}(\mathbf{R}^{n+1})$.

Q. E. D.

In order to work in a single Hilbert space we introduce the following unitary operator from $\mathcal{H}$ onto $\mathcal{H}_0$

$$(L\varphi)(x,y) = v^{-1}(x,y)\varphi(x,y), \tag{4.32}$$

where

$$v(x,y) = c(x,y)c_0^{-1}(y). \tag{4.33}$$

Under L, A_T is transformed into

$$\tilde{A}_T = L\, A_T\, L^{-1} = v\, A_{T,0}\, v, \tag{4.34}$$

where

$$D(\tilde{A}_T) = \{\varphi \in \mathcal{H}_0\ :\ v\varphi \in D(A_{T,0})\}. \tag{4.35}$$

We define the following selfadjoint nonnegative operators in $\mathcal{H}_0$

$$\tilde{A} = L\, A\, L^{-1} = v\, A_0 v, \tag{4.36}$$

$$D(\tilde{A}) = \{\varphi \in \mathcal{H}_0\ :\ v\varphi \in D(A_0)\}, \tag{4.37}$$

$$\tilde{A}_D = L\, A_{T,D}\, L^{-1} = v\, A_{D,0} v, \tag{4.38}$$

$$D(\tilde{A}_D) = \{\varphi \in \mathcal{H}_0\ :\ v\varphi \in D(A_{D,0})\}. \tag{4.39}$$

We denote for $M > 0$ such that $\tilde{A}_T + M > 0$

$$V = (\tilde{A}_T + M)^{-1} - (A_0 + M)^{-1}. \tag{4.40}$$

Lemma 4.5.

Suppose that Assumption 3.2 is satisfied, and that

$$|c(x, y) - c_0(y)| \leq C(1 + |x| + |y|)^{-\delta}, \tag{4.41}$$

for some constants $C,\ \delta > 0$.

Then the operator V extends to a bounded operator from $L^2_{-s_1}(\mathbf{R}^{n+1})$ into $L^2_{s_2}(\mathbf{R}^{n+1})$ for any $s_1 \geq 0,\ s_2 \geq 0$, with $s_1 + s_2 \leq \delta$, and it extends to a compact operator between the same spaces if $s_1 + s_2 < \delta$.

Proof: We find it convenient to decompose V in the following three terms

$$V = V_1 + V_2 + V_3, \tag{4.42}$$

where

$$V_1 = [(\tilde{A}_T + M)^{-1} - (\tilde{A}_D + M)^{-1}], \tag{4.43}$$

$$V_2 = -[(\tilde{A} + M)^{-1} - (\tilde{A}_D + M)^{-1}], \tag{4.44}$$

$$V_3 = [(\tilde{A} + M)^{-1} - (A_0 + M)^{-1}]. \tag{4.45}$$

Let $f(x,y) \in C_0^\infty(\mathbf{R}^{n+1})$ satisfy $f(x,y) \equiv 1$ in a neighborhood of $\partial\Omega$. Recall that the domains of A_T and Δ_T are the same set of functions. In consequence it follows from Assumption 3.2 that $f(A_T+M)^{-1}$ is a compact operator from $\mathcal{H}_0$ into $L^2_s(\mathbf{R}^{n+1})$ for any $s \geq 0$. Since

$$f(\tilde{A}_T + M)^{-1} = v^{-1} f(A_T + M)^{-1} v, \tag{4.46}$$

it follows that $f(\tilde{A}_T + M)^{-1}$ is compact from $\mathcal{H}_0$ into $L^2_s(\mathbf{R}^{n+1})$, for any $s \geq 0$.

Since $\tilde{A}_D$ and $\tilde{A}$ are particular cases of the operator $\tilde{A}_T$, it also follows that $f(\tilde{A}_D+M)^{-1}$ and $f(\tilde{A}+M)^{-1}$ are compact between the same spaces.

It follows that

$$f\, V = \sum_{i=1}^{3} f\, V_i, \tag{4.47}$$

is a compact operator from $\mathcal{H}_0$ into $L^2_s(\mathbf{R}^{n+1})$, for any $s \geq 0$.

Moreover,

$$\begin{aligned}(1-f)V_1 &= (1-f)[(A_{T,0}+M)^{-1} - (A_{D,0}+M)^{-1}] + \\ &+ (1-f)[(\tilde{A}_T+M)^{-1} - (A_{T,0}+M)^{-1}] - \\ &- (1-f)[(\tilde{A}_D+M)^{-1} - (A_{D,0}+M)^{-1}]. \end{aligned}\tag{4.48}$$

By Lemma 4.4 the first term in the right hand side of (4.48) is compact from $\mathcal{H}_0$ into $L^2_s(\mathbf{R}^{n+1})$, $\quad s \geq 0$. Moreover,

$$(\tilde{A}_T + M)^{-1} = v^{-1}(A_{T,0}+M)^{-1} Q(M) v^{-1}, \tag{4.49}$$

where

$$Q(M) = (A_{T,0}+M)v(\tilde{A}_T+M)^{-1}v \;=\; I + M(v-v^{-1})(\tilde{A}_T+M)^{-1}v. \tag{4.50}$$

Then

$$\begin{aligned}(\tilde{A}_T + M)^{-1} = v^{-1}(A_{T,0}+M)^{-1}v^{-1} &+ \\ + v^{-1}M(A_{T,0}+M)^{-1}(v-v^{-1})(\tilde{A}_T+M)^{-1}&. \end{aligned}\tag{4.51}$$

It follows that

$$\begin{aligned}&(1-f)[(\tilde{A}_T+M)^{-1} - (A_{T,0}+M)^{-1}] = \\ &\qquad (1-f)[(v^{-1}-1)(A_{T,0}+M)^{-1}v^{-1} + (A_{T,0}+M)^{-1}(v^{-1}-1) \\ &\qquad + v^{-1}M(A_{T,0}+M)^{-1}(v-v^{-1})(\tilde{A}_T+M)^{-1}]. \end{aligned}\tag{4.52}$$

It follows from Lemma 4.3 that the right hand side of (4.52) is bounded from $\mathcal{H}_0$ into $H_{2,s}(\mathbf{R}^{n+1})$, for $0 \leq s \leq \delta$, and since the imbedding of $H_{2,s_1}(\mathbf{R}^{n+1})$ into $L^2_{s_2}(\mathbf{R}^{n+1})$ is compact if $s_1 > s_2 \geq 0$, it is compact from $\mathcal{H}_0$ into $L^2_s(\mathbf{R}^{n+1})$ if $0 \leq s < \delta$.

Note that the third term in the right hand side of (4.48) has the same property because it is the particular case of the left hand side of (4.52) with Dirichlet boundary condition.

Then $(1-f)V_1$ is bounded from $\mathcal{H}_0$ into $L^2_s(\mathbf{R}^{n+1})$, for $0 \leq s \leq \delta$, and it is compact from $\mathcal{H}_0$ into $L^2_s(\mathbf{R}^{n+1})$, $\quad 0 \leq s < \delta$.

By a similar argument the operators $(1-f)V_i$, $\quad i = 2, 3$, have the same property. It follows that V also has the same property, and equivalently that

$$\eta_s(x,y)V \tag{4.53}$$

is bounded on $\mathcal{H}_0$ for $0 \leq s \leq \delta$, and that it is compact on $\mathcal{H}_0$ for $0 \leq s < \delta$. Then by taking adjoints and by Hadamard's three lines interpolation theorem (see Reed and Simon 1975, appendix to section IX.4) the operator

$$\eta_{s_2}(x,y)\; V \eta_{s_1}(x,y) \tag{4.54}$$

is bounded on $\mathcal{H}_0$ if $s_1 \geq 0$, $s_2 \geq 0$, $s_1 + s_2 \leq \delta$ and it is compact on $\mathcal{H}_0$ if $s_1 + s_2 < \delta$. This completes the proof of the lemma.

Q. E. D.

<u>**Lemma 4.6**</u>

Suppose that Assumption 3.2 is satisfied and that

$$|c(x,y) - c_0(y)| \leq C(1 + |x| + |y|)^{-\delta} \tag{4.55}$$

for some constants C, $\delta > 0$. Then the essential spectrum of A_T consists of $[0, \infty)$.

<u>**Proof**</u>: We proved in Theorem 1.1 that the spectrum of A_0 consists of $[0, \infty)$. Then also $\sigma_e(A_0) = [0, \infty)$. By Lemma 4.5 the difference of the resolvents of $\tilde{A}_T$ and A_0 is compact in $\mathcal{H}_0$. Then (see Schechter 1971, sections 1.4 and 1.7) $\tilde{A}_T$ and A_0 have the same essential spectrum. But since A_T and $\tilde{A}_T$ are unitarily equivalent

$$\sigma_e(A_T) = \sigma_e(\tilde{A}_T) = \sigma_e(A_0) \;=\; [0, \infty). \tag{4.56}$$

Q. E. D.

Note that since the perturbed propagator in all space, A, is the particular case of the transmission problem propagator, A_T, when $\overline{\Omega}_i$ is empty we have in particular that

$$\sigma_e(A) = [0, \infty). \tag{4.57}$$

Remark 4.7

We now show how the exterior domain case can also be put into the framework of the transmission problem. Given an exterior domain Ω_e, we denote $\Omega_i = \mathbf{R}^{n+1} \setminus \overline{\Omega}_e$ and $\Omega = \Omega_i \cup \Omega_e$.

The basic idea is to add to the given exterior domain problem an interior problem defined on Ω_i that has only a discrete spectrum, and that in consequence does not change the essential spectrum. In particular, we can take the interior Dirichlet problem.

We denote by $-\Delta_{i,D}$ the selfadjoint nonnegative operator in $L^2(\Omega_i)$ associated with the quadratic form (3.23) with domain $H_1^0(\Omega_i) \times H_1^0(\Omega_i)$. Since the domain of $\Delta_{i,D}$ is contained in $H_1^0(\Omega_i)$, it follows by the Rellich local compactness theorem that the resolvent of $A_{i,D}$ is compact and then $A_{i,D}$ has only discrete spectrum.

Note that

$$L^2(\mathbf{R}^{n+1}) = L^2(\Omega_i) \ \oplus \ L^2(\Omega_e). \tag{4.58}$$

We define

$$\Delta_T = \Delta_{i,D} \oplus \Delta_e, \tag{4.59}$$

where Δ_e is the given exterior domain Laplacian that satisfies Assumption 3.1. Δ_T is a selfadjoint bounded above extension of the Δ in $C_0^\infty(\Omega)$, and since

$$f(x,y)(-\Delta_T + M)^{-1} = f(x,y)(-\Delta_{i,D} + M)^{-1} \ \oplus \ f(x,y)(-\Delta_e + M)^{-1}, \tag{4.60}$$

Δ_T satisfies Assumption 3.2.

Let us suppose that $c(x,y)$ satisfies (4.55) on Ω_e, and let us extend it to $\mathbf{R}^{n+1}$ by taking $c(x,y) = 1$ on Ω_i (any extension compatible with (3.2) will do as well). We define A_T and $\tilde{A}_T$ as in (3.18) and (4.34) with Δ_T as in (4.59). Then by Lemma 4.6,

$$\sigma_e(\tilde{A}_T) = [0, \infty), \tag{4.61}$$

and since $\tilde{A}_T$ and A_T are unitarily equivalent,

$$\sigma_e(A_T) = [0, \infty). \tag{4.62}$$

Note that since $c(x, y) = 1$ on Ω_i

$$\mathcal{H} = L^2(\Omega_i) \oplus \mathcal{H}_e. \tag{4.63}$$

Moreover

$$A_T = -\Delta_{i,D} \oplus A_e, \tag{4.64}$$

and since $-\Delta_{i,D}$ has only discrete spectrum, it follows from (4.62) and (4.64) that

$$\sigma_e(A_e) = \sigma_e(A_T) = [0, \infty). \tag{4.65}$$

Q. E. D.

§5. Absence of Positive Eigenvalues

In the previous section we have proved that the essential spectrum of the perturbed acoustic propagator consists of $[0, \infty)$. We now study the positive eigenvalues imbedded in the essential spectrum. We will prove that for the transmission problem in Ω, the eigenfunctions corresponding to positive eigenvalues have their support contained in the complement of the unbounded component of Ω. In particular in the cases of the problem in all space and of connected exterior domains the perturbed acoustic propagator has no positive eigenvalues.

In the two theorems below we give conditions of the perturbation under which these statements are true. The techniques used in the proofs are quite different from each other. These two theorems give complementary conditions for the absence of positive eigenvalues.

We denote by $C^\infty(\mathbf{R}^{n+1})$ the space of all complex valued infinitely differentiable functions on $\mathbf{R}^{n+1}$.

Theorem 5.1

Let Ω_e be a connected exterior domain, that is to say a connected open set contained on $\mathbf{R}^{n+1}$ whose complement is compact. Let $c_0(y)$ be a real valued measurable function defined on $\mathbf{R}$, and $c(x, y)$ a real valued measurable funcion on Ω_e. Suppose that

$$0 < c_m \leq c_0(y) \leq c_M, \; y \in \mathbf{R}, \tag{5.1}$$

and that

$$0 < c_m \leq c(x,y) \leq c_M, \ (x,y) \in \Omega_e, \tag{5.2}$$

for some positive constants c_m and c_M. We assume that for some $\delta > 0$, $\alpha > 0$,

$$|c_0(y) - c_+| \leq C\ (1+y)^{-1-\delta}, \ y \geq 0, \tag{5.3}$$

$$|c_0(y) - c_-| \leq C\ (1+|y|)^{-1-\delta}, \ y \leq 0, \tag{5.4}$$

$$|c(x,y) - c_0| \leq C\ e^{-\alpha|x|}(1+|y|)^{\frac{-1-\delta}{2}}, \tag{5.5}$$

for $(x,y) \in \Omega_e$, and for some constants, c_+, c_-, and C, where $c_+ \leq c_-$.

For $M \in \mathbf{R}$ we denote

$$\mathbf{R}^{n+1}_M = \{(x,y) \in \mathbf{R}^{n+1} \ : \ y \geq M\}. \tag{5.6}$$

Suppose that $c_0(y)$ is a continuous function for $y \geq M$, where M is such that $\mathbf{R}^{n+1}_M \subset \Omega_e$, and that

$$c(x,y) = c_0(y) \geq M. \tag{5.7}$$

Let $\varphi(x,y) \in L^2(\Omega_e)$ be a solution in distribution sense of the equation in Ω_e,

$$-c^2(x,y)\Delta\varphi = \lambda\varphi, \tag{5.8}$$

for some $\lambda > 0$. Then $\varphi(x,y) = 0$ for a.e. $(x,y) \in \Omega_e$.

<u>**Proof**</u>: Suppose that $\Omega_i = \mathbf{R}^{n+1} \setminus \overline{\Omega}_e$ is contained in the ball of radius N, and let $f(x,y) \in C^\infty(\mathbf{R}^{n+1})$ satisfy, $0 \leq f(x,y) \leq 1$, $f(x,y) = 0$ in the ball of radius $N+1$, and $f(x,y) = 1$ on the complement of the ball of radius $N+2$. Let us extend $\varphi(x,y)$ to Ω_i by $\varphi(x,y) = 0$, $(x,y) \in \Omega_i$. Then

$$\psi(x,y) = f(x,y)\varphi(x,y) \in L^2(\mathbf{R}^{n+1}), \tag{5.9}$$

and satisfies in distribution sense on $\mathbf{R}^{n+1}$

$$-\Delta\psi - \lambda\ c_0^{-2}(y)\psi = g(x,y), \tag{5.10}$$

where

$$g(x,y) = -(\Delta f)\varphi - 2\ \nabla f \cdot \nabla\varphi + \lambda(c^{-2} - c_0^{-2})\psi, \tag{5.11}$$

and where we have extended $c(x,y)$ to Ω_i by $c(x,y) = 1$ for $(x,y) \in \Omega_i$.

Note that since Δf and ∇f have compact support on Ω_e, it follows from (5.8) and elliptic regularity that

$$e^{\alpha|x|}(1+|y|)^{\frac{1+\delta}{2}} g(x,y) \in L^2(\mathbf{R}^{n+1}). \tag{5.12}$$

Let $\mathcal{F}$ denote the Fourier transform in the x variables as defined in (1.11). We denote

$$\hat{\psi}(k,y) = (\mathcal{F}\psi)(k,y). \tag{5.13}$$

Since $\hat{\psi}(k,y) \in L^2(\mathbf{R}^{n+1})$ and because the test space $C_0^\infty(\mathbf{R}^{n+1})$ is separable, there is a set, O, of measure zero on $\mathbf{R}^n$ such that for $k \in \mathbf{R}^n \setminus O$

$$\hat{\psi}(k,y) \in L^2(\mathbf{R}), \tag{5.14}$$

and

$$\left(-\frac{d^2}{d_y^2} + \lambda\,(c_+^{-2} - c_0^{-2}(y)) - (\lambda\, c_+^{-2} - k^2) \right) \hat{\psi}(k,y) = \hat{g}(k,y), \tag{5.15}$$

where

$$\hat{g}(k,y) = (\mathcal{F}g)(k,y). \tag{5.16}$$

By (5.11) and (5.12) for almost every $x \in \mathbf{R}^n$

$$g(x,y) \in L_s^2(\mathbf{R}), \quad s = \frac{1+\delta}{2}, \tag{5.17}$$

and

$$\|g(x,\cdot)\|_{L_s^2(\mathbf{R})} \leq C\, e^{-\alpha|x|}. \tag{5.18}$$

Then we can consider $\hat{g}(k,y)$ as the Fourier transform of the function $x \to g(x,y)$ from $\mathbf{R}^n$ into $L_s^2(\mathbf{R})$, and it follows from (5.18) that

$$\hat{g}(k,y) \in L_s^2(\mathbf{R}) \tag{5.19}$$

for each fixed $k \in \mathbf{R}^n$, and that the function $k \to \hat{g}(k,y)$ is continuous from $\mathbf{R}^n$ into $L^2(\mathbf{R})$.

Moreover, let us take polar coordinates on $\mathbf{R}^n$, $k = \rho\nu$, $\rho > 0$, $\nu \in S_1^{n-1}$, and denote

$$\hat{g}(\rho,\nu,y) = \hat{g}(\rho\nu,y). \tag{5.20}$$

Then by (5.18) $\hat{g}(\rho,\nu,y)$ extends to a function of ρ and ν with values on $L_s^2(\mathbf{R})$, defined for $\rho \in \mathbf{C}$, $|Im\ \rho| < \alpha$, and $\nu \in S_1^{n-1}$, given by

$$\hat{g}(\rho,\nu,y) = \frac{1}{(2\pi)^{n/2}} \int_{\mathbf{R}^n} e^{-i\rho\nu\cdot x}\, g(x,y)dx. \tag{5.21}$$

Moreover $\hat{g}(\rho,\nu,y)$ is jointly continuous on ρ and ν, and for each fixed ν it is analytic on ρ with values on $L_s^2(\mathbf{R})$. In (5.21), $g(x,y)$ is integrated as a function with values on $L_s^2(\mathbf{R})$.

We will prove that $\hat{\psi}(k, y)$ extends to an analytic function of $|k|$. To do so we use the limiting absorption principle for the following operator in $L^2(\mathbf{R})$

$$h_1 = -\frac{d^2}{dy^2} + q_1(y), \tag{5.22}$$

$$D(h_1) = H_2(\mathbf{R}), \tag{5.23}$$

$$q_1(y) = \lambda\, c_+^{-2} - \lambda\, c_0^{-2}(y). \tag{5.24}$$

Note that

$$\lim_{y \to \infty} q_1(y) = 0, \tag{5.25}$$

$$\lim_{y \to -\infty} q_1(y) = q_- = \lambda\, c_+^{-2} - \lambda\, c_-^{-2} \geq 0. \tag{5.26}$$

Since $q_1(y)$ is bounded, h_1 is selfadjoint and bounded below. For $z \in \rho(h_1)$ we denote by

$$r_1(z) = (h_1 - z)^{-1}, \tag{5.27}$$

the resolvent of h_1. Note that for $z \in \rho(h_1)$, $r_1(z)$ a priori is an operator on $L^2(\mathbf{R})$, can be considered as an analytic operator valued function with values on $\mathcal{B}(L_s^2(\mathbf{R})\ ,\ H_{2,-s}(\mathbf{R}))$.

We prove in Appendix 2 that the following limits

$$r_1(\lambda \pm i\,0) = \lim_{\epsilon \downarrow 0} (h_1 - \lambda \mp i\,\epsilon)^{-1} \tag{5.28}$$

exist for $\lambda > 0$, in the uniform operator topology on $\mathcal{B}(L_s^2(\mathbf{R}), H_{2,-s}(\mathbf{R}))$, $s = \frac{1+\delta}{2}$.

Moreover the functions

$$r_{1,\pm}(z) = \begin{cases} r_1(z), & Im\ z \neq 0, \\ r_1(z \pm i\,0), & z \in (0, \infty), \end{cases} \tag{5.29}$$

defined for $z \in \mathbf{C}^{\pm} \cup (0, \infty)$ with values on $\mathcal{B}(L_s^2(\mathbf{R})\ ,\ H_{2,-s}(\mathbf{R}))$ are analytic for $Im\ z \neq 0$, and locally Hölder continuous for $z \in (0, \infty)$.

Note that $(-\infty, 0) \setminus \sigma_d(h_1)$ is contained in the resolvent set of h_1. Then we extend $r_{1,\pm}(z)$ to $(-\infty, 0) \setminus \sigma_d(h)$ as

$$r_{1,+}(z) = r_{1,-}(z) = r_1(z), \tag{5.30}$$

for $z \in (-\infty, 0) \setminus \sigma_d(h_1)$. Note that the common value on (5.30) is analytic. Consider now $r_{1,\pm}(\lambda\, c_+^{-2} - \rho^2)$ as functions of ρ defined for $\rho \in (D_\pm \cup (0,\infty) \setminus D_d)$, where

$$D_\pm = \{\rho \in \mathbf{C} \ : \ Re\ \rho > 0, \quad \mp Im\ \rho > 0\}, \tag{5.31}$$

$$D_d = \{\sqrt{\lambda\, c_+^{-2}}\} \bigcup_{i=1}^{\infty} \{\sqrt{\lambda\, c_+^{-2} - \lambda_i}\}, \tag{5.32}$$

where λ_i, $i = 1, 2, 3, \cdots$ are the negative eigenvalues of h_1. Note that h_1 has no positive eigenvalues (see Eastham and Kalf 1982).

Then $r_{1,\pm}(\lambda\, c_+^{-2} - \rho^2)$ are analytic functions of ρ, for $\rho \in D_\pm$, and are locally Hölder continuous for $\rho \in (0,\infty) \setminus D_d$, with values on $\mathcal{B}(L_s^2(\mathbf{R}), H_{2,-s}(\mathbf{R}))$.

Take any $k \in \mathbf{R}^n \setminus 0, \quad k \neq 0$. Then by (5.15), (5.20)

$$\lim_{\epsilon \downarrow 0} r_1(\lambda\, c_+^{-2} - k^2 \pm i\,\epsilon)\hat{g}\Big(|k|, \frac{k}{|k|}, y\Big)$$
$$= \hat{\psi}(k,y) \pm i\,\epsilon \lim_{\epsilon\downarrow 0} r_1(\lambda\, c_+^{-2} - k^2 \pm i\,\epsilon)\hat{\psi}(k,y). \tag{5.33}$$

Denote by $E_{h_1}(\mu)$ the family of spectral projectors of h_1. Then for any $\chi \in L^2(\mathbf{R})$ by the spectral theorem

$$\lim_{\epsilon\downarrow 0} \|\epsilon\, r_1(\beta \pm i\,\epsilon)\chi\|^2_{L^2(\mathbf{R})} = \lim_{\epsilon\downarrow 0} \int \frac{\epsilon^2}{(\mu-\beta)^2 + \epsilon^2}\, d(E_{h_1}(\mu)\chi, \chi) = 0, \tag{5.34}$$

provided that β is not an eigenvalue of h_1.

It follows from (5.33) and (5.34) that

$$\hat{\psi}(k,y) = r_{1,\pm}(\lambda\, c_+^{-2} - k^2)\hat{g}(|k|, \frac{k}{|k|}, y), \tag{5.35}$$

provided that $|k| \notin D_d$. Since O is of measure zero on $\mathbf{R}^n$ we can define $\hat{\psi}(k,y)$ as given by (5.35), for $k \in O$, and then (5.35) is true for all $k \in \mathbf{R}^n$ such that $|k| \notin D_d,\ k \neq O$.

Take again $k \in \mathbf{R}^n \setminus O, \quad k^2 < \lambda\, c_+^{-2}$. Then $\hat{\psi}(k,y) \in L^2(\mathbf{R})$ and (5.15) is valid. The homogeneous equation associated to (5.15) is

$$\Big(-\frac{d^2}{dy^2} + \lambda(c_+^{-2} - c_0^{-2}(y)) - (\lambda\, c_+^{-2} - k^2)\Big)v(k,y) = 0. \tag{5.36}$$

It follows from Example 3, Appendix 2, to Section 8 of Chapter XI of Reed and Simon 1979, that (5.36) has two solutions, $v_\pm(k,y)$, on $[M,\infty)$ such that $v_\pm(k,y)$ are continuously differentiable on y, and

$$\lim_{y\to\infty} \Big| v_\pm(k,y) - e^{(\pm\, i\sqrt{\lambda\, c_+^{-2} - k^2})y} \Big| = 0, \tag{5.37}$$

$$\lim_{y\to\infty} \Big| \frac{d}{dy} v_\pm(k,y) \mp i\sqrt{\lambda\, c_+^{-2} - k^2}\, e^{(\pm\, i\sqrt{\lambda\, c_+^{-2} - k^2})y} \Big| = 0. \tag{5.38}$$

Note that (by eventually taking a larger M) we can assume that $M > N+2$.

Since the right hand side of (5.15) is zero for $y \geq M$, $\hat{\psi}(k, y)$ is a solution to the homogeneous equation (5.36) on (M, ∞)

$$\left(-\frac{d^2}{dy^2} + \lambda\,(c_+^{-2} - c_0^{-2}(y)) - (\lambda\, c_+^{-2} - k^2)\right) \hat{\psi}(k, y) = 0, \tag{5.39}$$

and

$$\hat{\psi}(k, y) \in L^2(M, \infty). \tag{5.40}$$

Furthermore by (5.37), (5.38), $v_\pm(k, y)$ are a fundamental system of solutions to (5.36) on (M, ∞), and both are not on $L^2(M, \infty)$. It follows that for each fixed $k \in \mathbf{R}^n \setminus 0$

$$\hat{\psi}(k, y) = 0, \tag{5.41}$$

on (M, ∞). But by (5.35) $\hat{\psi}(k, y)$ is a continuous function of $k \in \mathbf{R}^n$, $|k| \neq 0$, $|k| \notin D_d$, into $H_{2,-s}(\mathbf{R}^n)$. Since (5.41) is valid for almost every $k \in \mathbf{R}^n$ with $k^2 < \lambda\, c_+^{-2}$, it follows from the continuity of $\hat{\psi}(k, y)$ that (5.41) holds for every $k \in \mathbf{R}^n$, $\quad 0 < k^2 < \lambda\, c_+^{-2}$.

We denote for $\rho > 0$, $\rho \notin D_d$, and $\omega \in S_1^{n-1}$

$$\hat{\psi}(\rho, \nu, y) = \hat{\psi}(\rho\nu, y). \tag{5.42}$$

Then by (5.35),

$$\hat{\psi}(\rho, \nu, y) = r_{1,\pm}(\lambda\, c_+^{-2} - \rho^2)\hat{g}(\rho, \nu, y). \tag{5.43}$$

Moreover, we denote

$$D_{\alpha,\pm} = \{D_\pm \cup (0, \infty) \setminus D_d\} \cap \{\rho \subset \mathbf{C}\ :\ |Im\ \rho| < \alpha\}, \tag{5.44}$$

and for $\rho \in D_{\alpha,\pm}$

$$\hat{\psi}_\pm(\rho, \nu, y) = r_{1,\pm}(\lambda\, c_+^{-2} - \rho^2)\hat{g}(\rho, \nu, y). \tag{5.45}$$

It follows from (5.35) and (5.43) that

$$\hat{\psi}_+(\rho, \nu, y) = \hat{\psi}_-(\rho, \nu, y) = \hat{\psi}(\rho, \nu, y), \tag{5.46}$$

for $\rho \in (0, \infty) \setminus D_d$.

Then $\hat{\psi}_\pm(\rho, \nu, y)$ define respectively an analytic continuation of $\hat{\psi}(\rho, \nu, y)$ to $D_\pm \cap \{\rho \in \mathbf{C}\ :\ |Im\ \rho|\ < \alpha\}$, and since by (5.46) both extensions coincide with $\hat{\psi}(\rho, \nu, y)$ for $\rho \in (0, \infty) \setminus D_d$, it follows that for each fixed $\nu \in S_1^{n-1}$, $\hat{\psi}(\rho, \nu, y)$ extends to an analytic function of ρ with values on

$H_{2,-s}(\mathbf{R})$ defined for $\rho \in D_{\alpha,+} \cup D_{\alpha,-}$. But by (5.41) for each fixed $\nu \in S_1^{n-1}$ and every ρ, with $0 < \rho < \lambda\, c_+^{-2}$

$$\chi_{(M,\infty)}(y)\hat{\psi}(\rho,\nu,y) = 0, \tag{5.47}$$

on $H_{2,-s}(\mathbf{R})$ where $\chi_{(M,\infty)}(y)$ is the characteristic function of (M,∞). Since the interval $(0, \lambda\, c_+^{-2})$ is inside the domain of analyticity of $\hat{\psi}(\rho,\nu,y)$, it follows from (5.47) that for each fixed $\nu \in S_1^{n-1}$

$$\chi_{(M,\infty)}(y)\hat{\psi}(\rho,\nu,y) = 0, \tag{5.48}$$

for all $\rho \in D_{\alpha,+} \cup D_{\alpha,-}$ and then (see (5.42)) by taking the inverse Fourier transform of $\hat{\psi}(k,y)$ we have

$$\psi(x,y) = 0, \qquad y \geq M, \tag{5.49}$$

and since $f(x,y) = 1$ for $y \geq M$, it follows from (5.9) that

$$\varphi(x,y) = 0, \qquad y \geq M. \tag{5.50}$$

But $\varphi(x,y)$ is a solution to (5.8) on Ω_e. Then by unique continuation (see Reed and Simon 1978, Appendix to Section 13 of Chapter XIII)

$$\varphi(x,y) = 0, \qquad \text{for a.e.} \qquad (x,y) \in \Omega_e. \tag{5.51}$$

Q. E. D.

As in section 4, let us denote by A, A_e, and A_T respectively the perturbed acoustic propagator on all space, exterior domain, and for the transmission problem. Then we have

Corollary 5.2

Suppose that the conditions of Theorem 5.1 are satisfied, then the perturbed acoustic propagator A on all space and A_e in a connected exterior domain have no positive eigenvalues. Moreover the eigenfunctions corresponding to positive eigenvalues of the perturbed acoustic propagator, A_T for the transmission problem in Ω have compact support contained in the complement of the unbounded component of Ω.

Proof: The corollary follows from Theorem 5.1. Note that in the case of the transmission problem the unbounded component of Ω is a connected exterior domain.

Q. E. D.

Remark 5.3

An important particular case of Theorem 5.1 is when $c_0(y)$ satisfies the condition of that theorem, and the perturbation is of compact support, that is to say when

$$c(x, y) = c_0(y), \tag{5.52}$$

for $|x| + |y| \geq N$, for some $N > 0$.

We state below a second result on the absence of positive eigenvalues of the perturbed acoustic propagator.

Theorem 5.4

Let Ω_e be a connected exterior domain contained on $\mathbf{R}^{n+1}$, and let $c_0(y)$ be a real valued measurable function defined on $\mathbf{R}$ that satisfies

$$0 < c_m \leq c_0(y) \leq c_M, \quad y \in \mathbf{R}, \tag{5.53}$$

for some positive constants c_m, c_M. Moreover, assume that

$$|c_0(y) - c_+| \leq C\ (1 + y)^{-1-\delta}, \quad y > 0, \tag{5.54}$$

$$|c_0(y) - c_-| \leq C\ (1 + |y|)^{-1-\delta}, \quad y \leq 0, \tag{5.55}$$

for some positive constants c_+, c_-, δ, and C, and where $c_+ \leq c_-$.

Let $c(x, y)$ be a real valued measurable function defined on Ω_e such that

$$0 < c_m \leq c(x, y) \leq c_M, \quad (x, y) \in \Omega_e, \tag{5.56}$$

and

$$|c(x, y) - c_0(y)| \leq C\ (1 + |x| + |y|)^{-1-\delta}, \tag{5.57}$$

$(x, y) \in \Omega_e$. Let $\varphi(x, y) \in L^2(\Omega_e)$ be a solution in distribution sense of the equation in Ω_e

$$-c^2(x, y)\Delta\varphi = \lambda\, \varphi, \tag{5.58}$$

for some $\lambda > 0$. Then $\varphi(x, y) = 0$ for a.e. $(x, y) \in \Omega_e$.

As in the case of Theorem 5.1 we have that

Corollary 5.5

Under the conditions of Theorem 5.4, the perturbed acoustic propagators on all space and in a connected exterior domain have no positive eigenvalues and the eigenfunctions corresponding to positive eigenvalues of the perturbed acoustic propagator of the transmission problem in Ω have compact support contained in the complement of the unbounded component of Ω.

Proof: The result follows from Theorem 5.4

Q. E. D.

Proof of Theorem 5.4: Let $\Omega_i = \mathbf{R}^{n+1} \setminus \overline{\Omega}_e$ be contained in the ball of radius N, and let $f(x,y) \in C^\infty(\mathbf{R}^{n+1})$ satisfy $0 \leq f(x,y) \leq 1$, $f(x,y) = 0$ on the ball of radius $N+1$, and $f(x,y) = 1$ on the complement of the ball of radius $N+2$. Denote

$$\psi(x,y) = f(x,y)\varphi(x,y) \in L^2(\mathbf{R}^{n+1}), \tag{5.59}$$

where we extended $\varphi(x,y)$ to Ω_i by $\varphi(x,y) = 0$, $(x,y) \in \Omega_i$.

We denote by H the following selfadjoint bounded below operator in $L^2(\mathbf{R}^{n+1})$

$$H = -\Delta + q_1(y) + q_2(x,y), \tag{5.60}$$

with domain, $D(H) = H_2(\mathbf{R}^{n+1})$, where $q_1(y)$ is as in (5.24), and

$$q_2(x,y) = \lambda\, c_0^{-2}(y) - \lambda\, c^{-2}(x,y), \tag{5.61}$$

and where we extended $c(x,y)$ to Ω_i by $c(x,y) = 1$, $(x,y) \in \Omega_i$.

Note that (5.25) and (5.26) are satisfied, and that by (5.57),

$$|q_2(x,y)| \leq C\, (1 + |x| + |y|)^{-1-\delta}. \tag{5.62}$$

It follows from (5.58) that

$$(H - \lambda\, c_+^{-2})\psi = b, \tag{5.63}$$

where

$$b(x,y) = (-\Delta f)\varphi - 2\nabla f \cdot \nabla\varphi \in L^2(\mathbf{R}^{n+1}). \tag{5.64}$$

Note that by elliptic regularity $\psi \in D(H)$, and that $b(x,y)$ has compact support contained in the ball of radius $N+2$.

For $\theta \in \mathbf{R}$ we denote by $U(\theta)$ the strongly continuous one parameter family of unitary operators on $L^2(\mathbf{R}^{n+1})$ given by

$$(U(\theta)\varphi)(x) = e^{-\frac{n+1}{2}\theta}\,\varphi(e^{-\theta}x,\ e^{-\theta}y), \tag{5.65}$$

for $\varphi \in L^2(\mathbf{R}^{n+1})$. $U(\theta)$ gives an unitary representation of the dilatation group. Let D denote the selfadjoint generator of $U(\theta)$ given by Stone's theorem (see Reed and Simon 1972, Chapter 8, Theorem VIII. 8),

$$U(\theta) = e^{-2i\theta D}. \tag{5.66}$$

Then it follows from (5.65) that

$$D = -\frac{i}{4}\left(\sum_{j=1}^{n}\left(x_j\,\frac{\partial}{\partial x_j} + \frac{\partial}{\partial x_j}\,x_j\right) + y\,\frac{\partial}{\partial y} + \frac{\partial}{\partial y}\,y\right), \tag{5.67}$$

defined on its maximal domain. It follows from Theorem XIII.10 on Chapter VIII of Reed and Simon 1972, that D is essentially selfadjoint on $C_0^\infty(\mathbf{R}^{n+1})$.

In what follows we denote by $(\cdot,\cdot)$ the scalar product on $L^2(\mathbf{R}^{n+1})$. We denote by $i[H,D]$ the following symmetric quadratic form with domain $D(H)\cap D(D)$

$$(i[H,D]\varphi,\psi) \;=\; i(D\varphi, H\psi) \;-\; i(H\varphi, D\psi). \tag{5.68}$$

Let $d(y) \in C^\infty(\mathbf{R})$ satisfy $d(y) = 1$ for $y \le 0$, and $d(y) = 0$, for $y \ge 1$.

Since both H and D are essentially selfadjoint on $C_0^\infty(\mathbf{R}^{n+1})$ it follows from a simple calculation that

$$(i[H,D]\varphi,\psi) = (-\Delta\varphi,\psi) + \frac{1}{2}\sum_{j=1}^{n}\Big[(\frac{\partial}{\partial x_j}\varphi, x_j\,q_2\psi) + (x_j\,q_2\varphi, \frac{\partial}{\partial x_j}\psi)\Big] +$$

$$+\frac{n}{2}(\varphi, q_2\psi) + \frac{1}{2}\Big[(\frac{\partial}{\partial y}\varphi, y(q_1 - dq_-)\psi) + (y(q_1 - dq_-)\varphi, \frac{\partial}{\partial y}\psi) +$$

$$+((q_1 - dq_-)\varphi,\psi)\Big] - \frac{1}{2}q_-((\frac{d}{dy}d)\varphi,\psi). \tag{5.69}$$

It follows from (5.69) that $i[H,D]$ is bounded below and closable. We denote also by $i[H,D]$ the selfadjoint bounded below operator associated with its closure (see Kato 1976, Chapter 6, Section 2).

The central issue in the proof of Theorem 5.4 is to establish that H satisfies an estimate introduced into spectral analysis by Mourre (see Mourre 1981, and Weder 1984b, where Mourre's estimate was first introduced into classical wave propagation).

The following Mourre estimate is proven in Weder 1988, Lemma 3.1: for any $\sigma > q_-$ there exists a non empty open interval Δ, that contains σ, a positive constant, γ, and a compact operator K such that

$$E(\Delta)i[H,D]\ E(\Delta) \geq \gamma\ E(\Delta) + K, \tag{5.70}$$

where $E(\cdot)$ denotes the spectral family of H.

Since the proof of (5.70) is rather technical and requires methods that are quite different from the main themes of this monograph we omit it here, and refer to Weder 1988 where a detailed proof is given.

Once estimate (5.70) is proved we proceed as in the proof of Theorem 2.1 of Froese and Herbst 1982. Note that since both the real and imaginary part of $\varphi(x,y)$ are a solution to equation (5.58) we can assume that $\varphi(x,y)$ is real valued, and then that $\psi(x,y)$ and $b(x,y)$ on (5.63) are also real valued. For $\delta > 0$, and $M > 0$, denote

$$F(x,y) = M\ \ell n\Big(\rho(x,y)(1+\delta\rho(x,y))^{-1}\Big),$$

where $\rho(x,y) = (1+|x|^2+|y|^2)^{\frac{1}{2}}$, and let $g(x,y)$ be such that $\nabla F = (xg, yg)$. We moreover denote $\psi_F = e^F\psi$, and by $H(F)$ the operator

$$H(F) = H - (\nabla\cdot F)^2 + \nabla\cdot\nabla\ F + \nabla F\cdot\nabla, \tag{5.71}$$

with domain

$$D(H(F)) = H_2(\mathbf{R}^{n+1}). \tag{5.72}$$

Then the following statements are proven as in Lemma 2.2 of Froese and Herbst 1982, where the case when $b(x,y) \equiv 0$ is considered

$$(H(F) - \lambda\ c_+^{-2})\psi_F = e^F b, \tag{5.73}$$

$$((H - \lambda\ c_+^{-2})\psi_F,\ \psi_F) = ((\nabla\cdot F)^2\psi_F + e^F b, \psi_F), \tag{5.74}$$

$$\begin{aligned}(\psi_F, i[H,D]\psi_F) \;&=\; -2\|g^{1/2}D\ \psi_F,\|^2 + \frac{1}{2}(\psi_F, \{(x\cdot\nabla)^2 g - \\ &- x\cdot\nabla(\nabla\cdot F)^2\}\psi_F) - 4\ Im\ (D\ \psi_F, e^F b),\end{aligned} \tag{5.75}$$

where by $\|\cdot\|$ we denote the norm on $L^2(\mathbf{R}^{n+1})$.

Moreover if we assume that for all $M > 0$ and for some fixed $\beta \geq 0$, $\rho^M e^{\beta\rho}\psi \in L^2(\mathbf{R}^{n+1})$, the statements above are also true when we take $F(x,y) = \beta\rho + M\ \ell n(1+\gamma\ \rho\ M^{-1})$, for all $M > 0$, and $\gamma > 0$.

We will prove that for all $\beta \geq 0$, $e^{\beta\rho(x,y)}\psi(x,y) \in L^2(\mathbf{R}^{n+1})$. Let us suppose first that for some $M > 0$, $\rho^M \psi(x,y) \notin L^2(\mathbf{R}^{n+1})$, and denote with $F = M\ \ell n\Big(\rho(x,y)(1+\epsilon\rho(x,y))^{-1}\Big)$, $\quad \epsilon > 0$

$$\psi_\epsilon(x,y) = \frac{e^F\psi}{\|e^F\psi\|}. \tag{5.76}$$

Note that if $\rho^M\psi(x,y) \notin L^2(\mathbf{R}^{n+1})$,

$$\lim_{\epsilon\downarrow 0} \|e^F\psi\| = \infty, \tag{5.77}$$

and it follows that

$$\lim_{\epsilon\downarrow 0} \left\| \frac{e^F b}{\|e^F\psi\|} \right\| = 0, \tag{5.78}$$

$$\lim_{\epsilon\downarrow 0} \left(\frac{e^F b}{\|e^F\psi\|}, \psi_\epsilon \right) = 0, \tag{5.79}$$

and

$$\lim_{\epsilon\downarrow 0} \left(D\ \psi_\epsilon, \frac{e^F b}{\|e^F\psi\|} \right) = 0. \tag{5.80}$$

It follows that the proof given in Froese and Herbst 1982, pages 435 and 436 applies and we obtain a contradiction. Then $\rho^M\psi(x,y) \in L^2(\mathbf{R}^{n+1})$ for all $M \geq 0$.

Denote

$$\beta_1 = \sup\{\alpha \geq 0\ :\ e^{\alpha\rho}\psi \in L^2(\mathbf{R}^{n+1})\}, \tag{5.81}$$

and assume that $\beta_1 < \infty$. Let Δ be as in (5.70) with $\sigma = \lambda\ c_+^{-2} + \beta_1^2$. If $\beta_1 = 0$ denote $\beta = \beta_1$, and if $\beta_1 > 0$, take β, $0 \leq \beta < \beta_1$, such that $\beta^2 + \lambda\ c_+^{-2} \in \Delta$. Take $\gamma > 0$ such that $\beta + \gamma > \beta_1$. We can always assume that $0 \leq \gamma \leq 1$. Take $F = \beta\rho + M\ \ell n(1 + \gamma\rho M^{-1})$, and denote

$$\psi_M = \frac{e^F\psi}{\|e^F\psi\|}. \tag{5.82}$$

Since

$$\lim_{M\to\infty} \|e^F\psi\| = \infty, \tag{5.83}$$

we have that

$$\lim_{M\to\infty} \left\| \frac{e^F b}{\|e^F\psi\|} \right\| = 0, \tag{5.84}$$

$$\lim_{M\to\infty} \left(\frac{e^F b}{\|e^F\psi\|}, \psi_M \right) = 0, \tag{5.85}$$

$$\lim_{M\to\infty} \left(D\ \psi_M, \frac{e^F b}{\|e^F\psi\|} \right) = 0. \tag{5.86}$$

It follows that the proof given in Froese and Herbst 1982, pages 436, 437, and 438 applies and we obtain a contradiction. In consequence $e^{\alpha\rho}\psi \in L^2(\mathbf{R}^{n+1})$ for all $\alpha \geq 0$.

We will prove now that $\psi(x, y)$ is of compact support. Let us suppose this is not the case and define

$$\psi_F = e^F \psi, \tag{5.87}$$

where $F = \alpha\rho$, and

$$\psi_\alpha = \frac{\psi_F}{\|\psi_F\|}. \tag{5.88}$$

Since formulas (5.73), (5.74) and (5.75) also hold with $F = \alpha\rho$, we have by (5.74) that

$$\|\nabla\, \psi_\alpha\|^2 = ((\lambda\, c_+^{-2} + \alpha^2)\psi_\alpha, \psi_\alpha) - (\rho^{-2}\psi_\alpha, \psi_\alpha) + \left(\frac{e^F b}{\|e^F \psi\|},\, \psi_\alpha\right). \tag{5.89}$$

Note that if $\psi(x, y)$ does not have compact support, for any $M > 0$,

$$\lim_{\alpha\to\infty} \int\limits_{|x|+|y|\leq M} |\psi_\alpha|^2 \, dxdy = 0. \tag{5.90}$$

Then

$$\lim_{\alpha\to\infty} (\rho^{-2}\psi_\alpha,\, \psi_\alpha) = 0. \tag{5.91}$$

Also if $\psi(x, y)$ does not have compact support

$$\lim_{\alpha\to\infty} \left\|\frac{e^F b}{\|\psi_F\|}\right\| = 0. \tag{5.92}$$

It follows from (5.89) that for some positive constant C_1,

$$\|\nabla\, \psi_\alpha\| \geq C_1 \alpha. \tag{5.93}$$

Moreover (5.54), (5.55), (5.57), and (5.69) imply that

$$(\psi_\alpha, i[H, D]\psi_\alpha) \geq \|\nabla\, \psi_\alpha\|^2 - C\, (\|\nabla\, \psi_\alpha\| + 1). \tag{5.94}$$

Also if $\psi(x, y)$ does not have compact support,

$$\lim_{\alpha\to\infty} \left(D\, \psi_\alpha,\, \frac{e^F b}{\|e^F \psi\|}\right) = 0. \tag{5.95}$$

Then by (5.54), (5.55), (5.57), (5.75), and (5.94)

$$\|\nabla\, \psi_\alpha\|^2 \leq C(\|\nabla\, \psi_\alpha\| + 1) + \frac{\alpha}{2}\,(\rho^{-1}\psi_\alpha, \psi_\alpha) + 4\left|\left(D\, \psi_\alpha,\, \frac{e^F b}{\|e^F \psi\|}\right)\right|. \tag{5.96}$$

But it follows from (5.93), (5.95), and (5.96), that for $\alpha \geq \alpha_0$, for some α_0 large enough,

$$C_1 \alpha \leq \|\nabla \psi_\alpha\| \leq \frac{C_2 \alpha}{(\|\nabla \psi_\alpha\| - C)} \leq C_3, \tag{5.97}$$

for some positive constants C, C_1, C_2, C_3. Then

$$\alpha \leq \frac{C_3}{C_1}, \tag{5.98}$$

for all $\alpha \geq \alpha_0$, and this is a contradiction.

It follows that $\psi(x, y)$ has compact support, and in consequence that $\varphi(x, y)$ has compact support on Ω_e. But $\varphi(x, y)$ is a solution to the equation

$$-c^2(x, y) \, \Delta \, \varphi = \lambda \varphi, \tag{5.99}$$

and it follows from unique continuation (see Reed and Simon 1978, appendix to Section 13 of Chapter XIII) that

$$\varphi(x, y) = 0, \tag{5.100}$$

for a.e. $(x, y) \in \Omega_e$.

Q. E. D.

§6. The Limiting Absorption Principle for the Perturbed Acoustic Propagator

In this section we will prove the limiting absorption principle for the perturbed acoustic propagator. We first consider the case of the transmission problem. The propagator in all space is a particular case. We then obtain the results for exterior domains from those for the transmission problem by adding a suitable interior domain.

Let A_T be the perturbed acoustic propagator for the transmission problem in Ω (see Section 3). We obtain the limiting absorption principle for A_T from the results already obtained for A_0 in Section 2. In order to work in a single Hilbert space, $\mathcal{H}_0$, we will consider the operator $\tilde{A}_T$ (see Section 4). The results for A_T will then follow since it is unitarily equivalent to $\tilde{A}_T$.

Let $M > 0$ be such that $\tilde{A}_T + M > 0$, and denote

$$H_0 = (A_0 + M)^{-1}, \tag{6.1}$$

$$H_1 = (\tilde{A}_T + M)^{-1}. \tag{6.2}$$

Then

$$H_1 = H_0 + V, \tag{6.3}$$

where

$$V = (\tilde{A}_T + M)^{-1} - (A_0 + M)^{-1}. \tag{6.4}$$

Recall that we proved in Lemma 4.5 that V has nice boundedness and compactness properties between appropriate weighted spaces. This suggests that we can obtain the limiting absorption principle for H_1 from that for H_0. Once the limiting absorption principle for H_1 is obtained, that for $\tilde{A}_T$ follows immediately.

We first obtain a spectral representation for H_0 in terms of that obtained for A_0 in Sections 1 and 2. Note that since the spectrum of A_0 is $[0, \infty)$, it follows from the functional calculus that that of H_0 is $[0, \frac{1}{M}]$, and that H_0 is absolutely continuous because A_0 is.

Let the operator F be defined as in Section 1. Then by (1.64) and functional calculus

$$F\, H_0 F^{-1} = (\lambda + M)^{-1}. \tag{6.5}$$

In order to obtain a spectral representation for H_0 we change the independent variable $\sigma = (\lambda + M)^{-1}$. For $\sigma \in \left(0, \frac{1}{M}\right)$, we define

$$\hat{\mathcal{H}}^0(\sigma) = \hat{\mathcal{H}}(\sigma^{-1} - M), \tag{6.6}$$

and

$$\hat{\mathcal{H}}^0(0) = \hat{\mathcal{H}}(\infty). \tag{6.7}$$

Moreover, we denote

$$\hat{\mathcal{H}}^0 = \oplus \int_{(0, M^{-1})} \hat{\mathcal{H}}^0(\sigma) d\sigma, \tag{6.8}$$

where $\hat{\mathcal{H}}(\lambda)$, $\lambda > 0$, is defined on (1.62).

Moreover, we denote by F^0 the following operator from $\mathcal{H}_0$ onto $\hat{\mathcal{H}}^0$,

$$(F^0 \varphi)(\sigma) = \frac{1}{\sigma}(F\varphi)(\sigma^{-1} - M). \tag{6.9}$$

It follows easily by changing variables from σ to $\lambda = \sigma^{-1} - M$, that F^0 is unitary from $\mathcal{H}_0$ onto $\hat{\mathcal{H}}^0$ and that

$$F^0\ H_0\ F^{0*} = \sigma, \tag{6.10}$$

the operator of multiplication by the independent variable on the direct integral (6.8).

We define for $\sigma \in (0, \frac{1}{M})$ and $s > 1/2$,

$$B^0(\sigma) = \frac{1}{\sigma}\, B(\sigma^{-1} - M), \tag{6.11}$$

where for $\lambda > 0$, $B(\lambda)$ is defined on (2.7). It then follows from the properties of $B(\lambda)$ established on Lemma 2.3 that $B^0(\sigma)$ is bounded from $\mathcal{H}_0$ into $\hat{\mathcal{H}}^0(\sigma) \subset \hat{\mathcal{H}}^0(0)$, and that the operator valued function $\sigma \longrightarrow B^0(\sigma)$ from $(0, \frac{1}{M})$ into $\mathcal{B}(\hat{\mathcal{H}}^0(0))$ is locally Hölder continuous with exponent γ that satisfies $\gamma < 1$, $\gamma < (s - 1/2)$ if $\sigma \neq \sigma_j$, $j = 1, 2, 3, \cdots$, and $\gamma < 1/2$, $\gamma < (s - 1/2)$ if $\sigma = \sigma_j$, for some $j = 1, 2, 3, \cdots$, where

$$\sigma_j = (\lambda_j + M)^{-1}, \tag{6.12}$$

with λ_j the thresholds (see 1.47).

Moreover by (6.10) we have that for every Borel set $\Delta \subset (0, \frac{1}{M})$, and $s > 1/2$

$$F^0 E^0(\Delta)\eta_{-s} = \chi_\Delta(\sigma) B^0(\sigma), \tag{6.13}$$

where $E^0(\cdot)$ denotes the family of spectral projectors of H_0.

For $z \in \mathbf{C} \setminus [0, \frac{1}{M}]$ we denote by $R^0(z)$ the resolvent of H_0

$$R^0(z) = (H_0 - z)^{-1}. \tag{6.14}$$

For any $\sigma \in (0, \frac{1}{M})$ denote by I_σ any compact interval, $I_\sigma \subset (0, \frac{1}{M})$ such that σ is an interior point of I_σ. Then for $z = \sigma \pm i\epsilon$, $\epsilon \neq 0$,

$$\eta_{-s} R^0(z) \eta_{-s} = \int_{I_\sigma} \frac{1}{\rho - z}\, B^{0*}(\rho) \quad B^0(\rho) d\rho + \eta_{-s} E^0(I_\sigma^\sim) R^0(z) \eta_{-s}, \tag{6.15}$$

for any $s > 1/2$, where we used (6.13), and we denote by $I_\sigma^\sim$ the complement of I_σ.

Then as in Theorem 2.4 we prove the following results: let us take $s > 1/2$. Then for every $\sigma > 0$,

$$R^0(\sigma \pm i0) = \lim_{\epsilon \downarrow 0} R^0(\sigma \pm i\epsilon) \tag{6.16}$$

exists in the uniform operator topology on $\mathcal{B}(L^2_s(\mathbf{R}^{n+1}), L^2_{-s}(\mathbf{R}^{n+1}))$, the convergence being uniform for σ in compact sets of $(0, \frac{1}{M})$. The following functions

$$R^0_\pm(\sigma) = \begin{cases} R^0(\sigma), & Im\ \sigma \neq 0, \\ R^0(\sigma \pm i0), & Im\ \sigma = 0, \end{cases} \tag{6.17}$$

defined for $\sigma \in \mathbf{C}^\pm \cup (0, \frac{1}{M})$ with values on $\mathcal{B}(L^2_s(\mathbf{R}^{n+1}), L^2_{-s}(\mathbf{R}^{n+1}))$ are analytic for $Im\ \sigma \neq 0$, and are locally Hölder continuous for $\sigma \in (0, \frac{1}{M})$, with exponent γ satisfying $\gamma < 1$, $\gamma < (s - 1/2)$ if $\sigma \neq \sigma_j$, $j = 1, 2, 3, \cdots$, and $\gamma < 1/2$, $\gamma < (s - 1/2)$, if $\sigma = \sigma_j$, for some $j = 1, 2, 3, \cdots$.

Furthermore, the $R^0_{\pm}(\sigma)$ are given by

$$\eta_{-s} R^0_{\pm}(\sigma) \eta_{-s} = P.\,V. \int_{I_\sigma} \frac{1}{\rho - \sigma} B^{0*}(\rho)\, B^0(\rho) d\rho \pm i\, \pi\, B^{0*}(\sigma) B^0(\sigma) +$$

$$+\ \eta_{-s}\, E^0(I_\sigma^{\sim}) R^0(\sigma) \eta_{-s}. \tag{6.18}$$

We proceed now to prove the limiting absorption principle for H_1. We assume that the local compactness assumption 3.2 holds and that

$$|c(x, y) - c_0(y)| \leq C\, (1 + |x| + |y|)^{-1-\delta}, \tag{6.19}$$

for some positive constants C, δ. By Lemma 4.5, V is compact on $\mathcal{H}_0$. Then since $\sigma(H_0) = \sigma_e(H_0) = [0, \frac{1}{M}]$, it follows that $\sigma_e(H_1) = [0, \frac{1}{M}]$.

For $\sigma \in \rho(H_1)$ we denote by

$$R^1(\sigma) = (H - \sigma)^{-1} \tag{6.20}$$

the resolvent of H_1. Then by (6.3)

$$R^1(\sigma) = R^0(\sigma) - R^1(\sigma) V\, R^0(\sigma), \tag{6.21}$$

and then

$$R^1(\sigma) Q_0(\sigma) = R^0(\sigma), \tag{6.22}$$

where

$$Q_0(\sigma) = (I + V\, R^0(\sigma)). \tag{6.23}$$

Note that $Q_0(\sigma)$ is invertible on $\mathcal{H}_0$ for $\sigma \in \rho(H_1)$ with inverse

$$Q(\sigma) = (H_0 - \sigma) R^1(\sigma). \tag{6.24}$$

The fact that $Q_0^{-1}(\sigma) = Q(\sigma)$ is immediate since by (6.23)

$$Q_0(\sigma) = (H_1 - \sigma) R^0(\sigma). \tag{6.25}$$

Then by (6.22) for $\sigma \in \rho(H_1)$

$$R^1(\sigma) = R^0(\sigma) Q_0^{-1}(\sigma) = R^0(\sigma) Q(\sigma). \tag{6.26}$$

The expression (6.26) suggests a direct way to extend $R^1(\sigma)$ to $(0, \frac{1}{M})$ from above and below.

Since for $\sigma \in \rho(H^1)$, both $R^1(\sigma)$ and $R^0(\sigma)$ are bounded operators on $\mathcal{H}_0$, they are also bounded operators from $L^2_s(\mathbf{R}^{n+1})$ into $L^2_{-s}(\mathbf{R}^{n+1})$, for any $s \geq 0$.

Then it follows from Lemma 4.5 and the limiting absorption principle for H_0 that the operators

$$V\, R^0_{\pm}(\sigma) \tag{6.27}$$

are compact on $L^2_s(\mathbf{R}^{n+1})$, for $s = \frac{1+\delta}{2}$ (note that the imbedding of $L^2_{s_1}(\mathbf{R}^{n+1})$, into $L^2_{s_2}(\mathbf{R}^{n+1})$ is bounded for $s_1 \geq s_2$).

Then $Q_0(\sigma)$ considered as a bounded operator on $L^2_s(\mathbf{R}^{n+1})$, $s = \frac{1+\delta}{2}$, has extensions to $\sigma \in \mathbf{C}^{\pm} \cup (0, \frac{1}{M})$ given by

$$Q_{0,\pm}(\sigma) = \begin{cases} Q_0(\sigma), & Im\, \sigma \neq 0, \\ I + V\, R^0_{\pm}(\sigma), & Im\, \sigma = 0. \end{cases} \tag{6.28}$$

The function $\sigma \to Q_{0,\pm}(\sigma)$ from $\mathbf{C}^{\pm} \cup (0, \frac{1}{M})$ into $\mathcal{B}(L^2_s(\mathbf{R}^{n+1}))$ are analytic for $\sigma \in \mathbf{C}^{\pm}$, and Hölder continuous for $\sigma \in (0, \frac{1}{M})$.

The expression (6.26) is equivalent to

$$R^1(\sigma) = R^1(\sigma) Q^{-1}_{0,\pm}(\sigma), \tag{6.29}$$

when both $R^1(\sigma)$ and $R^0(\sigma)$ are considered as bounded operators from $L^2_s(\mathbf{R}^{n+1})$ into $L^2_{-s}(\mathbf{R}^{n+1})$, $s = \frac{1+\delta}{2}$.

So far we only know that (6.29) is true for $\sigma \in \rho(H_1)$. However, if we prove that $Q_{0,\pm}(\sigma)$ is invertible for some $\sigma \in (0, \frac{1}{M})$, then (6.29) defines the extension to the point $\sigma \in (0, \frac{1}{M})$ from above or below that we are looking for.

From this point of view the limiting absorption principle for H_1 is formulated as the problem of studying the invertibility of $Q_{0,\pm}(\sigma)$ for $\sigma \in (0, \frac{1}{M})$. We proceed to analyse this question in the following lemma.

Lemma 6.1

Suppose that the local compactness assumption 3.2 is satisfied and that (6.19) holds. Then the operators $Q_{0,\pm}(\sigma)$ are invertible on $L^2_s(\mathbf{R}^{n+1})$, $s = \frac{1+\delta}{2}$, for $\sigma \in (0, \frac{1}{M})$ if and only if σ is not an eigenvalue of H_1.

Proof: Suppose that $\sigma \in (0, \frac{1}{M})$ is an eigenvalue H_1. Then for some $\varphi \in \mathcal{H}_0$, $\varphi \neq 0$,

$$(H_0 + V)\varphi = \sigma\varphi, \tag{6.30}$$

and

$$(H_0 - \sigma)\varphi = -V\, \varphi. \tag{6.31}$$

Then for all $\epsilon \neq 0$ we obtain by multiplying both sides of (6.31) by $R^0(\sigma \pm i\epsilon)$ that

$$\varphi \pm i\epsilon\, R^0(\sigma \pm i\,\epsilon)\varphi = -R^0(\sigma \pm i\epsilon)V\,\varphi. \tag{6.32}$$

But by the spectral theorem

$$\lim_{\epsilon\downarrow 0} \|\pm i\epsilon\, R^0(\sigma \pm i\epsilon)\varphi\|^2_{\mathcal{H}_0} = \lim_{\epsilon\downarrow 0} \int\limits_{(0,\frac{1}{M})} \frac{\epsilon^2}{(\rho-\sigma)^2+\epsilon^2}\, d(E^0(\rho)\varphi,\varphi) = 0, \tag{6.33}$$

since the spectral measure of H^0 is absolutely continuous.

By (6.32), (6.33), and the limiting absorption principle for H^0,

$$\varphi = -R^0_{\pm}(\sigma)V\,\varphi. \tag{6.34}$$

Note that since $\varphi \in \mathcal{H}_0$ it follows that $\psi = V\,\varphi \in L^2_s(\mathbf{R}^{n+1})$, $s = \frac{1+\delta}{2}$. Moreover, $\psi \neq 0$, because otherwise by (6.34) we would have that $\varphi = 0$. Then multiplying both sides of (6.34) by V we obtain that

$$\psi = -V\,R^0_{\pm}(\sigma)\psi, \tag{6.35}$$

and it follows that $Q_{0,\pm}(\sigma)$ is not invertible.

Suppose now that $Q_{0,\pm}(\sigma)$ is not invertible. Then for some $\psi_{\pm} \in L^2_s(\mathbf{R}^{n+1})$, $\psi_{\pm} \neq 0$

$$\psi_{\pm} = -V\,R^0_{\pm}(\sigma)\psi_{\pm}. \tag{6.36}$$

But $\varphi_{\pm} = R^0_{\pm}(\sigma)\psi_{\pm} \in L^2_{-s}(\mathbf{R}^{n+1})$, and $\varphi_{\pm} \neq 0$ because otherwise by (6.36) $\psi_{\pm} = 0$.

Multiplying both sides of (6.36) by $R^0_{\pm}(\sigma)$ we obtain that

$$\varphi_{\pm} = -R^0_{\pm}(\sigma)V\,\varphi_{\pm}. \tag{6.37}$$

We will prove below that $\varphi_{\pm}$ has a better decay rate at infinity than what we already know by the fact that it belongs to $L^2_{-s}(\mathbf{R}^{n+1})$. We will actually prove that $\varphi_{\pm} \in \mathcal{H}_0$.

Suppose that this has already been proved. Then by multiplying both sides of (6.37) by $(H_0 - \sigma)$ we obtain that

$$H_1\varphi_{\pm} = \sigma\varphi_{\pm}, \tag{6.38}$$

and it follows that σ is an eigenvalue of H_1.

To prove that $\varphi_{\pm} \in \mathcal{H}_0$ we multiply both sides of (6.37) by $\eta_{-s}(x,y)$, and by (6.18) we obtain that

$$\chi_{\pm} = -\Bigg[P.V. \int_{I_\sigma} \frac{1}{\rho-\sigma}\, B^{0*}(\rho)B^0(\rho)d\rho \pm i\pi\, B^{0*}(\sigma)B^0(\sigma) + \\ + \eta_{-s}E^0(I^{\sim}_\sigma)R^0(\sigma)\eta_{-s}\Bigg]\,\eta_s\,V\,\eta_s\,\chi_{\pm}, \tag{6.39}$$

where $\chi_\pm(x,y) = \eta_{-s}(x,y)\varphi_\pm(x,y) \in L^2(\mathbf{R}^{n+1})$. We will prove that in fact $\chi_\pm \in L^2_s(\mathbf{R}^{n+1})$, and in consequence that $\varphi_\pm \in \mathcal{H}_0$.

Note that by Lemma 4.5 $\eta_s\ V\ \eta_s$, $s = \frac{1+\delta}{2}$, is a compact operator on $L^2(\mathbf{R}^{n+1})$. It is also selfadjoint on $\mathcal{H}_0$ since V is. It follows from (6.39) that

$$\begin{aligned}\left(\chi_\pm, \eta_s\ V\ \eta_s\chi_\pm\right)_{\mathcal{H}_0} &= -P.V. \int_{I_\sigma} \frac{1}{\rho-\sigma} \left\|B^0(\rho)\eta_s\ V\ \eta_s\chi_\pm\right\|^2_{\mathcal{H}_0} \\ &\pm i\pi\left\|B^0(\sigma)\eta_s\ V\ \eta_s\chi_\pm\right\|^2_{\mathcal{H}_0} + \\ &+ ([\eta_{-s}E^0(I_\sigma^\sim)R^0(\sigma)\eta_{-s}]\eta_s\ V\ \eta_s\chi_\pm,\ \eta_s\ V\ \eta_s\chi_\pm).\end{aligned} \tag{6.40}$$

But since $\eta_s\ V\ \eta_s$ is selfadjoint on $\mathcal{H}_0$, the left hand side of (6.40) is real, and then the right hand side has also to be real. It follows that

$$B^0(\sigma)\eta_s\ V\ \eta_s\ \chi_\pm = 0, \tag{6.41}$$

where we used the fact that $\eta_{-s}E^0(I_\sigma^\sim)R^0(\sigma)\eta_{-s}$ is a selfadjoint operator on $\mathcal{H}_0$.

Equation (6.41) gives us a crucial piece of information because it implies (since $B^0(\rho)$ is locally Hölder continuous in ρ) that we can remove the principal value (*P.V.*) in (6.39) because

$$\frac{1}{\rho-\sigma}B^{0*}(\rho)B^0(\rho)\eta_s\ V\ \eta_s\ \chi_\pm = \frac{1}{\rho-\sigma}B^{0*}(\rho)(B^0(\rho)-B^0(\sigma))\eta_s\ V\ \eta_s\ \chi_\pm \tag{6.42}$$

is a locally integrable function of ρ. Then

$$\chi_\pm = \chi_{1,\pm} + \chi_{2,\pm}, \tag{6.43}$$

where

$$\chi_{1,\pm} = \int_{I_\sigma} \frac{1}{\rho-\sigma}B^{0*}(\rho)B^0(\rho)\eta_s\ V\ \eta_s\ \chi_\pm d\rho, \tag{6.44}$$

$$\chi_{2,\pm} = [\eta_{-s}E^0(I_\sigma^\sim)R^0(\sigma)\eta_{-s}]\eta_s\ V\ \eta_s\ \chi_\pm. \tag{6.45}$$

But since $E^0(I_\sigma^\sim)R^0(\sigma)$ is a bounded operator on $\mathcal{H}_0$ we have that

$$\left\|\chi_{2,\pm}\right\|_{L^2_s(\mathbf{R}^{n+1})} \leq C\left\|\chi_\pm\right\|_{\mathcal{H}_0}, \tag{6.46}$$

and it follows that $\chi_{2,\pm} \in L^2_s(\mathbf{R}^{n+1})$, $s = \frac{1+\delta}{2}$.

We will now prove that also $\chi_{1,\pm} \in L^2_s(\mathbf{R}^{n+1})$. We denote

$$\hat{\mathcal{H}}^0_\sigma = \oplus \int_{I_\sigma} \hat{\mathcal{H}}^0(\rho)d\rho. \tag{6.47}$$

Note that there is a $k > 0$ such that $(0, \sigma_j) \cap I_\sigma$ is not empty for all $j \leq k$, and that $(0, \sigma_j) \cap I_\sigma$ is empty for all $j \geq k$. Then we can identify

$$\hat{\mathcal{H}}^0_\sigma = L^2(I_\sigma,\ L^2(S_c)) \bigoplus_{j=1}^{k} L^2(I_{\sigma,j},\ L^2(S_1^{n-1})), \tag{6.48}$$

where

$$I_{\sigma,j} = (0, \sigma_j) \cap I_\sigma.$$

We denote by J_σ the following operator defined for $\varphi(\rho) \in \hat{\mathcal{H}}^0_\sigma$ by

$$J_\sigma(\varphi) = \int_{I_\sigma} B^{0*}(\rho)\varphi(\rho)d\rho. \tag{6.49}$$

It follows from (6.13) that

$$J_\sigma(\varphi) = \eta_{-s} E^0(I_\sigma) F^{0*} \varphi, \tag{6.50}$$

where we have extended φ to $I_\sigma^\sim$ by zero.

Since F^0 is unitary we have that

$$\left\| J_\sigma(\varphi) \right\|_{L^2_s(\mathbf{R}^{n+1})} \leq C \left\| E^0(I_\sigma) F^{0*} \varphi \right\| = C \left\| \varphi \right\|_{\hat{\mathcal{H}}^0_\sigma}. \tag{6.51}$$

Then $J_\sigma \in \mathcal{B}(\hat{\mathcal{H}}^0_\sigma,\ L^2_s(\mathbf{R}^{n+1}))$.

Moreover, note that if

$$\varphi(\rho) = \bigoplus_{j=0}^{k} \varphi_j(\rho), \tag{6.52}$$

$J_\sigma(\varphi)$ is given by

$$J_\sigma(\varphi) = \int_{I_\sigma} \frac{1}{\rho}\ \eta_{-s} T_0^*(\rho^{-1} - M)\varphi_0(\rho)d\rho + \sum_{j=1}^{k} \frac{1}{\rho}\ T_j^*(\rho^{-1} - M)\varphi_j(\rho)d\rho, \tag{6.53}$$

where we used (2.7) and (6.11). It follows from (6.53) that J_σ is also a bounded operator from

$$L^1(I_\sigma,\ L^2(S_c)) \bigoplus_{j=1}^{k} L^1(I_{\sigma,j},\ L^2(S_1^{n-1})), \tag{6.54}$$

into $\mathcal{H}_0$.

Then by interpolation (see Lions and Magenes 1972, and Reed and Simon 1975, Appendix to Section IX.4), J_σ is a bounded operator from

$$L^p(I_\sigma,\ L^2(S_c)) \bigoplus_{j=0}^{k} L^p(I_{\sigma,j},\ L^2(S_1^{n-1})), \tag{6.55}$$

into $L^2_{\epsilon_p s}(\mathbf{R}^{n+1})$, where $\epsilon_p = 2(1-\frac{1}{p})$, $1 \le p \le 2$.

Remark that by (6.44),

$$\chi_{1,\pm} = -J_\sigma(\frac{1}{\rho-\sigma}\ B^0(\rho)\eta_s\ V\ \eta_s\,\chi_\pm). \tag{6.56}$$

Suppose that we already know that $\chi_\pm \in L^2_{s_1}(\mathbf{R}^{n+1})$ for some $0 \le s_1 < s$. We can write (6.56) as (see (6.41))

$$\chi_{1,\pm} = -J_\sigma\left(\frac{1}{\rho-\sigma}(B^0(\rho)-B^0(\sigma))\eta_{-s_1}(\eta_{s_1}\eta_s\ V\ \eta_s\eta_{-s_1})\eta_{s_1}\chi_\pm\right). \tag{6.57}$$

Assume moreover that $\sigma \neq \sigma_j, \quad j = 1,2,3$. Then

$$\|(B^0(\rho)-B^0(\sigma))\eta_{-s_1}\|_{\mathcal{B}(\mathcal{H}_0,\hat{\mathcal{H}}^0(0))} \le C\,|\rho-\sigma|^\beta, \tag{6.58}$$

where $\beta < \min(1,\ s+s_1-\frac{1}{2})$, and it follows that

$$\frac{1}{\rho-\sigma}(B^0(\rho)-B^0(\sigma))\eta_{-s_1}(\eta_{s_1}\eta_s\ V\ \eta_s\eta_{-s_1})\eta_{s_1}\chi_\pm \in$$

$$\in L^p(I_\sigma,\ L^2(S_c)) \bigoplus_{i=1}^{k} L^p(I_{\sigma,j},\ L^2(S_1^{n-1})), \tag{6.59}$$

for any $p < (1-\beta)^{-1}$. But then by (6.57) $\chi_{1,\pm} \in L^2_{\epsilon_p s}(\mathbf{R}^{n+1})$, with $\epsilon_p = 2(1-\frac{1}{p})$, and $p \le 2, \quad p < (1-\beta)^{-1}$. Since $\chi_{2,\pm} \in L^2_s(\mathbf{R}^{n+1})$ it follows that $\chi_\pm \in L^2_{\epsilon_p s}(\mathbf{R}^{n+1})$.

We now start this argument with $s_1 = 0$, and we obtain that $\chi_\pm \in L^2_{s^1}(\mathbf{R}^{n+1})$, $s^1 = \epsilon_p s$, for any $p \le 2$, $p < (3/2-s)^{-1}$. We iterate the argument a finite number of times until we prove that (6.58) holds with some $p \ge 2$. Then we iterate once more to obtain that $\chi_{1,\pm} \in L^2_s(\mathbf{R}^{n+1})$, and in consequence that $\chi_\pm \in L^2_s(\mathbf{R}^{n+1})$, and finally that $\varphi_\pm \in \mathcal{H}_0$ as desired.

Consider now the case when $\sigma = \sigma_j$ for some $j = 1,2,3$. In this case (6.58) holds but with $\beta < \frac{1}{2}$, $\beta < (s+s_1-\frac{1}{2})$, and by iterating the argument above as before we only obtain that (6.59) holds with any $p < 2$ (recall that $p < (1-\beta)^{-1}$). In consequence with the argument above we only prove that $\chi_\pm \in L^2_{\tilde{s}}(\mathbf{R}^{n+1})$, $0 < \tilde{s} < s$, and since $\varphi_\pm = \eta_s\ \chi_\pm$, that

$$\varphi_\pm \in L^2_{-s'}(\mathbf{R}^{n+1}), \tag{6.60}$$

$0 < s' < 1/2$. So in this case we do not yet conclude that $\varphi_\pm \in \mathcal{H}_0$. However, the fact that $s' < 1/2$ in (6.60) allows us to write the equation for $\varphi_\pm$ in terms of an associated operator whose generalized eigenfunctions have better regularity properties than those of H_0, as we show below.

Since by Lemma 4.1 $(A_0 + M)^{-1}$ is bounded from $L^2_s(\mathbf{R}^{n+1})$, into $L^2_s(\mathbf{R}^{n+1})$ for $s \geq 0$, it follows by taking adjoints that it is also bounded from $L^2_{-s}(\mathbf{R}^{n+1})$ into $L^2_{-s}(\mathbf{R}^{n+1})$, $s \geq 0$. Then we can multiply both sides of (6.37) by $H_0 - \sigma$ to obtain

$$(H_0 - \sigma)\varphi_\pm = -V\ \varphi_\pm. \tag{6.61}$$

Let $f(x) \in C_0^\infty(\mathbf{R}^{n+1})$, $0 \leq f \leq 1$, $f(x) = 1$, $|x| \leq 1$, and $f(x) = 0$, $|x| \geq 2$, and denote $f_\ell(x,y) = f(\frac{x}{\ell}, \frac{y}{\ell})$. For any $\varphi \in D(\tilde{A}_T)$ with compact support

$$\begin{aligned}((H_0 - \sigma)\varphi_\pm, (\tilde{A}_T + M)\varphi)_{\mathcal{H}_0} &= -\lim_{\ell \to \infty} (V\ f_\ell\varphi_\pm, (\tilde{A}_T + M)\varphi)_{\mathcal{H}_0} = \\ &= (H_0\varphi_\pm, (\tilde{A}_T + M)\varphi)_{\mathcal{H}_0} - (\varphi_\pm, \varphi)_{\mathcal{H}_0}.\end{aligned} \tag{6.62}$$

It follows that

$$(\varphi_\pm, \tilde{A}_T\varphi)_{\mathcal{H}_0} = \lambda(\varphi_\pm, \varphi)_{\mathcal{H}_0}, \tag{6.63}$$

where $\lambda = \sigma^{-1} - M > 0$, and then that

$$(\tilde{A}_T\varphi_\pm, \varphi)_{\mathcal{H}_0} = \lambda(\varphi_\pm, \varphi)_{\mathcal{H}_0}. \tag{6.64}$$

In consequence

$$\tilde{A}_T\varphi_\pm = \lambda\ \varphi_\pm, \tag{6.65}$$

in distribution sense on Ω. Denote $\Sigma_\pm = v(x,y)\varphi_\pm$. Then (6.65) is equivalent to

$$-c^2\ \Delta_T\Sigma_\pm = \lambda\ \Sigma_\pm. \tag{6.66}$$

Let Ω_i be contained in the ball of radius N, and let $f(x,y) \in C^\infty(\mathbf{R}^{n+1})$ satisfy $f(x,y) = 0$, in the ball of radius $N+1$, and $f(x,y) = 1$ on the complement of the ball of radius $N+2$. Denote

$$\Gamma_\pm = f(x,y)\Sigma_\pm, \tag{6.67}$$

where we extended $\Sigma_\pm$ to Ω_i by $\Sigma_\pm(x,y) = 0$, $(x,y) \in \Omega_i$.

It follows from (6.66) that $\Gamma_\pm$ satisfies

$$(B - \lambda\ c_+^{-2})\Gamma_\pm = g_\pm, \tag{6.68}$$

where

$$g_{\pm}(x,y) = -(\Delta f)\Sigma_{\pm} - 2\nabla f \cdot \nabla \Sigma_{\pm} + \lambda(c^{-2} - c_0^{-2})\Gamma_{\pm}, \tag{6.69}$$

and

$$B = -\Delta + q(y), \tag{6.70}$$

$$D(B) = H_2(\mathbf{R}^{n+1}), \tag{6.71}$$

$$q(y) = -\lambda\, c_0^{-2}(y) + \lambda\, c_+^{-2}. \tag{6.72}$$

Note that by elliptic regularity and (6.19) $g(x,y) \in L^2_{1+\delta-s'}(\mathbf{R}^{n+1})$, and that $1+\delta-s' > 1/2$. In Appendix 2 we prove that limiting absorption principle holds for B.

Let $\varphi \in L^2_{-s'}(\mathbf{R}^{n+1})$, with $s' < 1/2$. Then for any $\gamma > 1/2$ (see Appendix 2 where we prove that for $z \in \mathbf{C} \setminus [0,\infty)$, $(B-z)^{-1}$ is a bounded operator from $L^2_s(\mathbf{R}^{n+1})$ into $L^2_s(\mathbf{R}^{n+1})$ for any $s \in \mathbf{R}$)

$$\begin{aligned} &s - \lim_{\epsilon\downarrow 0} \epsilon\, \eta_{-\gamma}(B - \lambda\, c_+^{-2} \mp i\,\epsilon)^{-1}\varphi = \\ &= s - \lim_{\epsilon\downarrow 0} \Big[\eta_{-\gamma}\eta_{s'}\epsilon(B - \lambda\, c_+^{-2})^{-1}\eta_{-s'}\varphi + \eta_{-\gamma}(B - \lambda\, c_+^{-2} \mp i\,\epsilon)^{-1} \\ &\cdot [\eta_{s'}, B]\epsilon\, (B - \lambda\, c_+^{-2} \mp i\,\epsilon)^{-1}\eta_{-s'}\varphi\Big] = 0, \end{aligned} \tag{6.73}$$

where we used the facts that by the limiting absorption principle for B (see Appendix 2),

$$\left\|\eta_{-\gamma}(B - \lambda\, c_+^{-2} \mp i\,\epsilon)^{-1}\eta_{-(1-s')}\right\|_{\mathcal{B}(L^2(\mathbf{R}^{n+1}))} \le C, \tag{6.74}$$

for a constant C independent of ϵ (note that since $s' < 1/2$, $1 - s' > 1/2$), and since B has no eigenvalues (see Appendix 2) it follows from the spectral theorem with $D(B) = H_2(\mathbf{R}^{n+1})$ that

$$\lim_{\epsilon\downarrow 0} \epsilon \left\|(B - \lambda\, c_+^{-2} \mp i\,\epsilon)^{-1}\eta_{-s'}\varphi\right\|_{H_2(\mathbf{R}^{n+1})} = 0, \tag{6.75}$$

and where the limits on (6.73) are in the strong topology on $L^2(\mathbf{R}^{n+1})$. It follows that if $\varphi \in L^2_{s'}(\mathbf{R}^{n+1})$, with $s' < 1/2$

$$s - \lim_{\epsilon\downarrow 0} \epsilon(B - \lambda\, c_+^{-2} \mp i\,\epsilon)^{-1}\varphi = 0, \tag{6.76}$$

in the strong topology on $L^2_{-\gamma}(\mathbf{R}^{n+1})$, for any $\gamma > 1/2$.

It follows from (6.68), and (6.76) since $\Gamma_{\pm} \in L^2_{-s'}(\mathbf{R}^{n+1})$, $0 < s' < 1/2$, that

$$\Gamma_{\pm} = R_B(\lambda + i\, 0) g_{\pm}, \tag{6.77}$$

$$\Gamma_{\pm} = R_B(\lambda - i\, 0) g_{\pm}, \tag{6.78}$$

where $R_B(\lambda \pm i\, 0)$ are the extended resolvents of B (see Appendix 2). Note that in order to conclude that (6.77), (6.78) follow from (6.68) it is essential to know that $\Gamma_{\pm} \in L^2_{-s'}(\mathbf{R}^{n+1})$, with $s' < 1/2$. If we only know that $\Gamma_{\pm} \in L^2_{-s}(\mathbf{R}^{n+1})$ with $s > 1/2$, (6.68) does not in general imply (6.77) or (6.78).

Since both (6.77) and (6.78) hold we must have that (see Appendix 2)

$$C_+(\sqrt{\lambda\, c_+^{-2} - q_-})\eta_s g_{\pm} \;=\; C_-(\sqrt{\lambda\, c_+^{-2} - q_-})\eta_s g_{\pm} \;=$$

$$= C_0(\sqrt{\lambda\, c_+^{-2} - q_-})\eta_s g_{\pm} = 0. \tag{6.79}$$

$$C_j(\sqrt{\lambda\, c_+^{-2} - \lambda_j})\eta_s g_{\pm} = 0, \quad j = 1, 2, \cdots, Q, \tag{6.80}$$

where $s = \frac{1+\delta}{2}$. We denote

$$\tau_{\pm} = \eta_{-s}\Gamma_{\pm}. \tag{6.81}$$

Then it follows from (6.77) or (6.78) that (see Appendix 2)

$$\tau_{\pm} = \tau_{1,\pm} + \tau_{2,\pm}, \tag{6.82}$$

where

$$\tau_{1,\pm} = \left[\int_{I_{\lambda\, c_+^{-2}}} \frac{d\rho}{\rho^2 + q_- - \lambda\, c_+^{-2}} \left[C_+^*(\rho)C_+(\rho) + C_-^*(\rho)C_-(\rho) + \right.\right.$$

$$\left.\left. + C_0^*(\rho)C_0(\rho) \right] + \sum_{j=1}^{Q} \int_{I_j} \frac{d\rho}{\rho^2 + \lambda_j - \lambda\, c_+^{-2}}\, C_j^*(\rho)C_j(\rho) \right] \cdot \eta_s g_{\pm}, \tag{6.83}$$

$$\tau_{2,\pm} \;=\; \eta_{-s} R_B(\lambda\, c_+^{-2}) P_B([a,b]^{\sim}) g_{\pm}(x,y). \tag{6.84}$$

Then since $R_B(\lambda\, c_+^{-2})P_B([a,b]^{\sim})$ is a bounded operator on $\mathcal{H}_0$ we have that

$$\left\| \tau_{2,\pm} \right\|_{L^2_s(\mathbf{R}^{n+1})} \le C \left\| g_{\pm} \right\|_{\mathcal{H}_0}. \tag{6.85}$$

Moreover (6.83) is equivalent to (see Appendix 2)

$$\tau_{1,\pm} = M_0\Bigg(\frac{1}{\rho^2 + q_- - \lambda\, c_+^{-2}} \Big[C_+(\rho) - C_+(\sqrt{\lambda\, c_+^{-2} - q_-}) +$$

$$+C_-(\rho) - C_-(\sqrt{\lambda\, c_+^{-2} - q_-}) \; + \; C_0(\rho) - C_0(\sqrt{\lambda\, c_+^{-2} - q_-})\Big]\eta_s g_\pm \Bigg) +$$

$$+\sum_{j=1}^{Q} M_j\Bigg(\frac{1}{\rho^2 + \lambda_j - \lambda c_+^{-2}} \Big[C_j(\rho) - C_j(\sqrt{\lambda\, c_+^{-2} - \lambda_j})\Big]\eta_s g_\pm \Bigg), \quad (6.86)$$

where the operator $M_0 \in \mathcal{B}\Big(L^p(I_{\lambda c_+^{-2}},\, L^2(S_c)), L^2_{\epsilon_p s}(\mathbf{R}^{n+1})\Big)$ and $M_j \in \mathcal{B}\Big(L^p(I_j,\, L^2(S_1^{n-1})),\quad L^2_{\epsilon_p s}(\mathbf{R}^{n+1})\Big)$, for $1 \leq p \leq 2$, and where $\epsilon_p = 2\,(1 - \frac{1}{p})$.

The point now is that the operator valued functions $\rho \to C_\pm(\rho)$, $\rho \to C_j(\rho), j = 0, 1, 2, 3, \cdots$, are locally Hölder continuous on $(0, \infty)$ with exponent γ that satisfies $\gamma < 1,\ \gamma < (s - 1/2)$ (see Appendix 2). Then if we know that $\tau_\pm \in L^2_{\tilde{s}}(\mathbf{R}^{n+1}),\ 0 \leq \tilde{s} < s$, it follows that

$$\frac{1}{\rho^2 + q_- - \lambda\, c_+^{-2}} \Big[\Big(C_+(\rho) - C_+(\sqrt{q_- - \lambda\, c_+^{-2}}) +$$

$$+ C_-(\rho) - C_-(\sqrt{q_- - \lambda\, c_+^{-2}}) + C_0(\rho) -$$

$$- C_0(\sqrt{q_- - \lambda\, c_+^{-2}})\Big)\eta_{-\tilde{s}}\Big] \; \eta_{\tilde{s}}\eta_s g_\pm \in L^p(I_{\lambda\, c_+^{-2}},\, L^2(S_c)), \quad (6.87)$$

$$\frac{1}{\rho^2 + \lambda_j - \lambda\, c_+^{-2}} \Big[\Big(C_j(\rho) - C_j(\lambda\, c_+^{-2} - \lambda_j)\Big)\eta_{-\tilde{s}}\Big] \; \eta_s \eta_{\tilde{s}} g_\pm$$

$$\in L^p(I_j,\, L^2(S_1^{n-1})), \quad (6.88)$$

for any $1 \leq p < (1 - \beta)^{-1}$, where $\beta < \min\,(1,\ s + \tilde{s} - \frac{1}{2})$. Then by (6.85) and (6.86) $\tau_\pm \in L^2_{\epsilon_p s}(\mathbf{R}^{n+1})$. We start this argument with $\tilde{s} = s - s' > 0$, and we iterate until we prove that $\tau_\pm \in L^2_{\tilde{s}}(\mathbf{R}^{n+1})$. Then it follows by (6.67) and (6.81) that $\varphi_\pm = v^{-1}\, \Sigma_\pm \in \mathcal{H}_0$, and as before that $\sigma \in \sigma_p(H_1)$.

Q. E. D.

We proceed now to prove the limiting absorption principle for A_T. For $z \in \rho(A_T)$ we denote the resolvent of A_T by

$$R_T(z) = (A_T - z)^{-1}. \quad (6.89)$$

Note that since $R_T(z) \in \mathcal{B}(\mathcal{H}_0)$ we have that $R_T(z) \in \mathcal{B}(L^2_s(\mathbf{R}^{n+1}),\ L^2_{-s}(\mathbf{R}^{n+1}))$ for any $s \geq 0$. By $\sigma_+(A_T)$ we denote the set of positive eigenvalues of A_T.

Theorem 6.2

Suppose that the local compactness Assumption 3.2 is satisfied and that

$$|c(x,y) - c_0(y)| \leq C\ (1 + |x| + |y|)^{-1-\delta}, \tag{6.90}$$

for some constants $C,\ \delta > 0$.

Then the essential spectrum of A_T consists of $[0, \infty)$ and A_T has no singular continuous spectrum. The positive eigenvalues of A_T have finite multiplicity and can only accumulate at zero and infinite. The eigenfunctions corresponding to positive eigenvalues of A_T have their support contained in the complement of the unbounded component of Ω. For every $\lambda \in \mathbf{R} \setminus \sigma_+(A_T)$,

$$R_T(\lambda\ \pm\ i\ 0) = \lim_{\epsilon \downarrow 0} R_T(\lambda\ \pm\ i\ \epsilon), \tag{6.91}$$

exist in the uniform operator topology on $\mathcal{B}(L^2_s(\mathbf{R}^{n+1}),\ L^2_{-s}(\mathbf{R}^{n+1})),\ s = \frac{1+\delta}{2}$.
Furthermore the limit is uniform for λ in compact sets of $(0, \infty) \setminus \sigma_+(A_T)$.

Moreover the functions

$$R_{T,\pm}(\lambda) = \begin{cases} R_T(\lambda), & Im\ \lambda \neq 0, \\ R_T(\lambda\ \pm\ i\ 0), & Im\ \lambda = 0, \end{cases} \tag{6.92}$$

defined for $\lambda \in \mathbf{C}^{\pm} \cup [(0,\infty) \setminus \sigma_+(A_T)]$ with values on $\mathcal{B}(L^2_s(\mathbf{R}^{n+1}),\ L^2_{-s}(\mathbf{R}^{n+1})),\ s = \frac{1+\delta}{2}$, are analytic for $\lambda \in \mathbf{C}^{\pm}$, and locally Hölder continuous for $\lambda \in (0,\infty) \setminus \sigma_+(A_T)$ with exponent γ satisfying $\gamma < 1,\ \gamma < (s - 1/2)$, if $\lambda \neq \lambda_j,\quad j = 1, 2, 3, \cdots$, and $\gamma < 1/2,\ \gamma < (s - 1/2)$, if $\lambda = \lambda_j$, for some $j = 1, 2, 3, \cdots$.

Proof: The fact that $\sigma_e(A_T) = [0, \infty)$ has already been proved in Lemma 4.6. We proved in Corollary 5.5 that the eigenfunctions corresponding to the positive eigenvalues of A_T have their support contained on the complement of the unbounded component of Ω.

To prove that the positive eigenvalues of A_T have finite multiplicity and can accumulate only at zero and infinite assumes on the contrary that there is a sequence β_j of positive eigenvalues with an associated orthonormal sequence of eigenvectors φ_j and where

$$\lim_{j\to\infty} \beta_j = \beta, \tag{6.93}$$

for some $0 < \beta < \infty$.

Suppose that the complement of the unbounded component of Ω is contained in the ball of radius N, and let $f(x, y) \in C_0^\infty(\mathbf{R}^{n+1})$ satisfy $f(x, y) \equiv 1$ on the ball of radius $N + 1$. Then

$$\varphi_j = f\varphi_j = (\beta_j + M)\ f(A_T + M)^{-1}\varphi_j. \tag{6.94}$$

But by the local compactness assumption 3.2, $f(A_T + M)^{-1}$ is a compact operator (note that Δ_T and A_T have the same domain). Then by (6.94) the φ_j have a convergent subsequence, but this is impossible since the φ_j are an orthonormal sequence.

To prove the existence of the limits in (6.91) we first consider the case of H_1. We define

$$R^1_\pm(\sigma) = \begin{cases} R^1(\sigma), & Im\ \sigma \neq 0, \\ R^0_\pm(\sigma\ \pm\ i\ 0) & Q^{-1}_{0,\pm}(\sigma), \end{cases} \tag{6.95}$$

for $\sigma \in \mathbf{C}^\pm \cup [(0, M^{-1}) \setminus \sigma_+(H_1)]$, with values on $\mathcal{B}(L^2_s(\mathbf{R}^{n+1}), L^2_{-s}(\mathbf{R}^{n+1}))$, $s = \frac{1+\delta}{2}$, where $\sigma_+(H_1)$ denotes the set of positive eigenvalues of H_1, and where we used Lemma 6.1. Then (see the stability of bounded invertibility Theorem 1.16, Chapter IV, Section 1 of Kato 1976) the functions $\sigma \to R^1_\pm(\sigma)$ are analytic for $\sigma \in \mathbf{C}^\pm$ and locally Hölder continuous for $\sigma \in (0, \frac{1}{M}) \setminus \sigma_+(H_1)$ with the same exponent as $R^0_\pm(\sigma)$.

As the reader can easily check for $\lambda \in \rho(A_T) \equiv \rho(\tilde{A}_T)$,

$$(\tilde{A}_T - \lambda)^{-1} = -(\lambda + M)^{-1}\ [I + (\lambda + M)^{-1}R^1((\lambda + M)^{-1})]. \tag{6.96}$$

Then by (4.34)

$$(A_T - \lambda)^{-1} = L^{-1}\ (\tilde{A}_T - \lambda)^{-1}\ L = - v(\lambda + M)^{-1}[I + (\lambda + M)^{-1}R^1((\lambda + M)^{-1})]v^{-1} \tag{6.97}$$

It follows that the existence of the limits on (6.91) satisfying the properties stated in the theorem follow from (6.96) and the results already proven for $R^1_\pm(\sigma)$.

Finally the absence of singular continuous spectrum follows from the existence of the limits on (6.91) uniformly for λ in compact sets of $(0, \infty)$ (see Reed and Simon 1978, Section XIII.6)

Q. E. D.

Remark 6.3

Note that it follows from (6.96) that the limits on (6.91) are given by

$$R_T(\lambda \pm i\,0) = -v(\lambda+M)^{-1}[I+(\lambda+M)^{-1}\,R^1_{\mp}((\lambda+M)^{-1})]v^{-1}, \quad (6.98)$$

with $R^1_{\mp}((\lambda+M)^{-1})$ as in (6.95).

We now study the dependence of the extended resolvents on $c(x,y)$. We first define

Definition 6.4

Let $c(x,y)$ be a real valued measurable function on $\mathbf{R}^{n+1}$ that satisfies (3.2) and (6.90). A neighborhood, O_c, of c is defined as the set of all real valued measurable functions, $d(x,y)$, on $\mathbf{R}^{n+1}$ that satisfy (3.2) and (6.90) with the same constants c_m, c_M, and δ as $c(x,y)$, and such that for some fixed $\epsilon > 0$

$$\|\eta_{1+\delta}(c-d)\|_\infty \leq \epsilon, \quad (6.99)$$

for all $d \in O_c$.

We now make explicit the dependence on $c(x,y)$ of the operators and denote by $A_{T,c}$, $\tilde{A}_{T,c}$, $A_{T,d}$, $\tilde{A}_{T,d}$ the operators A_T and $\tilde{A}_T$ corresponding respectively to $c(x,y)$ and $d(x,y)$. We also denote

$$V_c = (\tilde{A}_{T,c}+M)^{-1} - (A_0+M)^{-1}, \quad (6.100)$$

$$V_d = (\tilde{A}_{T,d}+M)^{-1} - (A_0+M)^{-1}. \quad (6.101)$$

Note that we can choose the same M for all $d \in O_c$.

Lemma 6.5

There is a neighborhood, O_c, of c such that

$$\left\|V_d - V_c\right\|_{\mathcal{B}(L^2_{-s}(\mathbf{R}^{n+1}),\ L^2_s(\mathbf{R}^{n+1}))} \leq C\left\|\eta_{1+\delta}(d-c)\right\|_\infty, \quad (6.102)$$

for a fixed constant C and all $d \in O_c$, and where $s = \frac{1+\delta}{2}$.

Proof: Note that (see (4.49))

$$(\tilde{A}_{T,c}+M)^{-1} = v^{-1}\,(A_{T,0}+M)^{-1}\,Q_c(M)v^{-1}, \tag{6.103}$$

where

$$Q_c(M) = (A_{T,0}+M)v\,(\tilde{A}_{T,c}+M)^{-1}\,v. \tag{6.104}$$

Denote

$$Q_{0,c}(M) = v^{-1}\,(\tilde{A}_{T,c}+M)v^{-1}\,(A_{T,0}+M)^{-1}. \tag{6.105}$$

Note that

$$Q_{0,c}(M) = I + M(v^{-2}-1)\,(A_{T,0}+M)^{-1}, \tag{6.106}$$

and that

$$Q_c(M) = (Q_{0,c}(M))^{-1}. \tag{6.107}$$

Then by (6.103)

$$(\tilde{A}_{T,c}+M)^{-1} = v^{-1}(A_{T,0}+M)^{-1}\,(I+M(v^{-2}-1)\,(A_{T,0}+M)^{-1})^{-1}v^{-1}. \tag{6.108}$$

By stability of bounded invertibility (see Theorem 1.16, Chapter IV, Section 1 of Kato 1976) there is a $\epsilon > 0$ such that

$$\left\|(\tilde{A}_{T,d}+M)^{-1} - (\tilde{A}_{T,c}+M)^{-1}\right\|_{\mathcal{B}(\mathcal{H}_0)} \leq C\,\left\|d-c\right\|_{\infty}, \tag{6.109}$$

for a fixed constant C and all $d(x,y)$ with

$$\|d-c\|_\infty \leq \epsilon. \tag{6.110}$$

Moreover we have that (see (4.51))

$$(\tilde{A}_{T,c}+M)^{-1} = v^{-1}\,(A_{T,0}+M)^{-1}\,v^{-1} + v^{-1}\,M(A_{T,0}+M)^{-1}$$

$$\cdot\;(v-v^{-1})\,(\tilde{A}_{T,c}+M)^{-1}, \tag{6.111}$$

and

$$(\tilde{A}_{T,d}+M)^{-1} = v_1^{-1}\,(A_{T,0}+M)^{-1}v_1^{-1} + v_1^{-1}\,M(A_{T,0}+M)^{-1}$$

$$\cdot\;(v_1-v_1^{-1})\,(A_{T,d}+M)^{-1}, \tag{6.112}$$

where $v_1(x,y) = d(x,y)\, c_0^{-1}(y)$.

It follows that

$$V_d - V_c = (v_1^{-1} - v^{-1})\,(A_{T,0} + M)^{-1}\, v^{-1} + (v_1^{-1} - 1)\,(A_{T,0} + M)^{-1}$$

$$\cdot (v_1^{-1} - v^{-1}) + (A_{T,0} + M)^{-1}\,(v_1^{-1} - v^{-1}) - M\,(v_1^{-1} - v^{-1})$$

$$\cdot (A_{T,0} + M)^{-1}(v^{-1} - v)\,(\tilde{A}_{T,c} + M)^{-1} - M v_1^{-1}\,(A_{T,0} + M)^{-1}$$

$$\cdot ((v_1^{-1} - v^{-1}) - (v_1 - v))\,(\tilde{A}_{T,c} + M)^{-1} - M v_1^{-1}\,(A_{T,0} + M)^{-1}\,(v_1^{-1} - v)$$

$$\cdot ((\tilde{A}_{T,d} + M)^{-1} - (\tilde{A}_{T,c} + M)^{-1}). \tag{6.113}$$

By Lemma 4.3 $(A_{T,0} + M)^{-1}$ is a bounded operator from $L_s^2(\mathbf{R}^{n+1})$ into $L_s^2(\mathbf{R}^{n+1})$ for any $s \geq 0$.

It follows from (6.109) and (6.113) that there is a neighborhood, O_c, of c such that

$$\left\| \eta_{1+\delta}\,(V_d - V_c) \right\|_{\mathcal{B}(\mathcal{H}_0)} \leq C\, \left\| \eta_{1+\delta}\,(c-d) \right\|_{\infty}, \tag{6.114}$$

for all $d \in O_c$, and some constant C.

By taking adjoints

$$\left\| (V_d - V_c)\eta_{1+\delta} \right\|_{\mathcal{B}(\mathcal{H}_0)} \leq C\, \left\| \eta_{1+\delta}\,(c-d) \right\|_{\infty}, \tag{6.115}$$

for all $d \in O_c$, then by Hadamard's three lines interpolation theorem (see Reed and Simon 1975, Appendix to Section 4 of Chapter IX),

$$\left\| \eta_s\,(V_d - V_c)\eta_s \right\|_{\mathcal{B}(\mathcal{H}_0)} \leq C\, \left\| \eta_{1+\delta}\,(c-d) \right\|_{\infty}, \tag{6.116}$$

$s = \frac{1+\delta}{2}$, and the Lemma follows.

Q. E. D.

We will denote by $R_{T,\pm}^c(\lambda)$ the extended resolvents of $A_{T,c}$.

Lemma 6.6

The extended resolvents $R_{T,\pm}^c(\lambda)$ are locally Lipschitz continuous on $c(x,y)$ uniformly for λ in compact sets of $(0,\infty) \setminus \sigma_+(A_{T,c})$: for each compact set $K \subset (0,\infty) \setminus \sigma_+(A_{T,c})$ there exists a neighborhood, O_c, of $c(x,y)$ such that for all $d(x,y) \in O_c$, $\quad K \subset (0,\infty) \setminus \sigma_+(A_{T,d})$ and

$$\left\| R^c_{T,\pm}(\lambda) \; - \; R^d_{T,\pm}(\lambda) \right\|_{\mathcal{B}(L^2_s(\mathbf{R}^{n+1}),\; L^2_{-s}(\mathbf{R}^{n+1}))}$$

$$\leq C \left\| (1+|x|+|y|)^{1+\delta} \, (c-d) \right\|_{\infty}, \tag{6.117}$$

for some constant C and all $\lambda \in K$, $\quad d \in O_c$, and where $s = \frac{1+\delta}{2}$.

Proof: We denote by

$$\tilde{K} = \{\sigma \in (0, M^{-1}) \; : \; \sigma = (\lambda + M)^{-1}, \quad \text{for some} \quad \lambda \in K\}. \tag{6.118}$$

It follows by functional calculus that $K \subset (0, \frac{1}{M}) \setminus \sigma_+(H_{1,c})$, where $H_{1,c} = (\tilde{A}_{T,c} + M)^{-1}$. Note that since $A_{T,c}$ and $\tilde{A}_{T,c}$ are unitarily equivalent they have the same spectrum.

By Lemmas 6.1 and 6.5 and by the stability of bounded invertibility (see Theorem 1.16, Chapter IV, Section 1 of Kato 1976) there is a neighborhood, O_c, of $c(x,y)$, such that for all $\sigma \in \tilde{K}$, $\quad I + V_d R^0_{\pm}(\sigma)$ is invertible for all $d \in O_c$ and $\sigma \in \tilde{K}$. Then $\tilde{K} \in (0, \frac{1}{M}) \setminus \sigma_+(H_{1,d})$ for all $d \in O_c$, and it follows from functional calculus that $K \subset (0,\infty) \setminus \sigma_+(A_{T,d})$, for all $d \in O_c$.

Moreover, it follows from (6.95) and stability of bounded invertibility that for all $\sigma \in \tilde{K}$ and $d \in O_c$ (by $R^{1,c}_{\pm}(\sigma)$ we denote $R^1_{\pm}(\sigma)$ for $H_{1,c}$),

$$\left\| R^{1,c}_{\pm}(\sigma) \; - \; R^{1,d}_{\pm}(\sigma) \right\|_{\mathcal{B}(L^2_s(\mathbf{R}^{n+1}),\; L^2_{-s}(\mathbf{R}^{n+1}))}$$

$$\leq C \left\| (1+|x|+|y|)^{1+\delta} \, (c-d) \right\|_{\infty}, \tag{6.119}$$

for some constant C. The estimate (6.117) follows now from (6.98) and (6.119).

Q. E. D.

Remark 6.7

In the particular case when $\overline{\Omega}_i = \emptyset$ we obtain the propagator on full space $\mathbf{R}^{n+1}$. In this case we know by Corollary 5.5 that A has no positive eigenvalues. Moreover, zero is obviously not an eigenvalue because the only harmonic function on $\mathbf{R}^{n+1}$ that is square integrable in the zero function. It then follows from Theorem 6.2 that A is absolutely continuous.

Moreover for all $z \in \rho(\tilde{A}_T)$

$$(\tilde{A} - z)^{-1} = v^{-1} R_0(z) G(z) v^{-1}, \tag{6.120}$$

where

$$G(z) = (A_0 - z) v (\tilde{A} - z)^{-1} v. \tag{6.121}$$

Denote

$$G_0(z) = v^{-1}(\tilde{A} - z) v^{-1} \, R_0(z) = I + z(1 - v^{-2}) R_0(z). \tag{6.122}$$

Then both $G(z)$ and $G_0(z)$ are invertible for $z \in \rho(\tilde{A}_T)$ and

$$G(z) = (G_0(z))^{-1}. \tag{6.123}$$

It follows from (6.120) that

$$(\tilde{A} - z)^{-1} = v^{-1} \, R_0(z) G_0^{-1}(z) v^{-1}. \tag{6.124}$$

Define for $\lambda \in \mathbf{C}^{\pm} \cup (0, \infty)$

$$G_{0,\pm}(\lambda) = \begin{cases} G_0(\lambda), & Im \, \lambda \neq 0, \\ I + \lambda \, (1 - v^{-2}) \, R_{0,\pm}(\lambda), & \lambda \in (0, \infty). \end{cases} \tag{6.125}$$

Then by (4.36) for $\lambda \in \mathbf{C}^{\pm}$,

$$R(\lambda) = R_0(\lambda) \; G_0^{-1}(\lambda) v^{-2}. \tag{6.126}$$

Moreover, we have that

$$R_{\pm}(\lambda) = R_{0,\pm}(\lambda) \; (I + \lambda(1 - v^{-2}) \; R_{0,\pm}(\lambda))^{-1} v^{-2}, \tag{6.127}$$

for $\lambda \in \mathbf{C}^{\pm} \cup (0, \infty)$. A simple way to see this is as follows: let us prove that $(I + \lambda(1 - v^{-2}) R_{0,\pm}(\lambda))$ is invertible on $L^2_s(\mathbf{R}^{n+1})$, $s = \frac{1+\delta}{2}$. Suppose that this is not true. Then since $(1 - v^{-2}) R_{0,\pm}(\lambda)$ is compact on $L^2(\mathbf{R}^{n+1})$ (note that $R_{0,\pm}(\lambda)$ is bounded from $L^2_{s'}(\mathbf{R}^{n+1})$ into $H_{2,-s'}(\mathbf{R}^{n+1})$ for any $s' > 1/2$), there exists $\varphi \in L^2_s(\mathbf{R}^{n+1})$, $\varphi \neq 0$, such that

$$G_{0,\pm}(\lambda) \varphi = 0. \tag{6.128}$$

But by (6.126) for any $\epsilon > 0$,

$$R(\lambda \, \pm \, i \, \epsilon) v^2 \; G_{0,\pm}(\lambda \, \pm \, i \, \epsilon) \varphi = R_0(\lambda \, \pm \, i \, \epsilon) \varphi, \tag{6.129}$$

and by taking the limit when $\epsilon \longrightarrow 0$,

$$0 = R(\lambda \pm i\,0)v^{-2}\; G_{0,\pm}(\lambda)\varphi = R_{0,\pm}(\lambda)\varphi. \tag{6.130}$$

But since

$$(H_0 - \lambda)\; R_{0,\pm}(\lambda)\varphi = \varphi, \tag{6.131}$$

It follows that $\varphi = 0$. Then $(I + \lambda(1 - v^{-2})R_{0,\pm}(\lambda))$ is invertible for all $\lambda \in (0, \infty)$, and (6.127) follows from (6.126) by taking the limits on the real axis from above and below.

Note that since

$$D(A) = H_2(\mathbf{R}^{n+1}), \tag{6.132}$$

it follows as in the proof of Theorem 2.4 that we can replace in Theorem 6.2 and in Lemma 6.6 the uniform operator topology in $\mathcal{B}(L^2_s(\mathbf{R}^{n+1}),\; L^2_{-s}(\mathbf{R}^{n+1}))$, by the uniform operator topology in $\mathcal{B}(L^2_s(\mathbf{R}^{n+1}),\; H_{2,-s}(\mathbf{R}^{n+1}))$.

Romark 6.8

Let us consider the exterior domain problem in Ω_e, which can be put into the framework of the transmission problem as in Remark 4.7. However we find it convenient to proceed in a slightly more general way.

We denote $\Omega_i = \mathbf{R}^{n+1} \setminus \overline{\Omega}_e$, and $\Omega = \Omega_i \cup \Omega_e$. Let us denote by $\Delta_{i,D}$, the Dirichlet Laplacian in $L^2(\Omega_i)$ as in Remark 4.7.

We define

$$\Delta_T = (\Delta_{i,D} - N)\; \oplus\; \Delta_e, \tag{6.133}$$

where $N \geq 0$. As in Remark 4.7, Assumption 3.2 is satisfied if Assumption 3.1 holds. We extend $c(x, y)$ to Ω_i by $c(x, y) = 1$, and define

$$A_T = -c^2(x, y)\Delta_T, \tag{6.134}$$

with Δ_T as in (6.133).

We denote by $\mathcal{H}_{i,0}$ the Hilbert space of all complex valued square integrable functions on Ω_i with scalar product

$$(\varphi, \psi)_{\mathcal{H}_{i,0}} = \int_{\Omega_i} \varphi(x, y)\; \overline{\psi}(x, y)\; c_0^{-2}(y)dxdy. \tag{6.135}$$

Remark that

$$\mathcal{H}_0 = \mathcal{H}_{i,0}\; \oplus\; \mathcal{H}_{e,0}, \tag{6.136}$$

and $\tilde{A}_T = c_0^2(y)\; (-\Delta_{i,D} + N)\; \oplus\; \tilde{A}_e$.

Note that Lemma 4.3 holds for Δ_T as in (6.133) (the proof only needs a slight modification). Also Lemma 4.4 holds since

$$(A_{T,0}+M)^{-1}-(A_{D,0}+M)^{-1} =$$

$$= \left[\left((c^{-2}(-\Delta_{i,D}+N)+M)^{-1}-(-c_0^2\,\Delta_{i,D}+M)^{-1}\right)\oplus 0\right]+$$

$$+\left[((-c_0^2\,\Delta_i+M)^{-1}\oplus(-c_0^2\,\Delta_e+M)^{-1}) - (A_{D,0}+M)^{-1}\right]. \tag{6.137}$$

By the Rellich local compactness theorem the operator

$$\left[\left((c_0^2(-\Delta_{i,D}+N)+M)^{-1}-(-c_0^2\,\Delta_{i,D}+M)^{-1}\right)\oplus 0\right] \tag{6.138}$$

is compact from $\mathcal{H}_0$ into $L_s^2(\mathbf{R}^{n+1})$ for any $s\geq 0$. By Lemma 4.4 the operator

$$[\,((-c_0^2\,\Delta_i+M)^{-1}\oplus(-c_0^2\,\Delta_e+M)^{-1})-(A_{D,0}+M)^{-1}], \tag{6.139}$$

is compact from $\mathcal{H}_0$ onto $L_s^2(\mathbf{R}^{n+1})$, for any $s\geq 0$. Then by taking adjoints and by interpolation it follows as in the proof of Lemma 4.4 that

$$(A_{T,0}+M)^{-1}-(A_{D,0}+M)^{-1} \tag{6.140}$$

is compact from $L_{-s_1}^2(\mathbf{R}^{n+1})$ into $L_{s_2}^2(\mathbf{R}^{n+1})$, for any $s_1,\ s_2\geq 0$.

Also Lemma 4.5 holds with A_T as in (6.134). Then if we assume that (6.90) holds, Theorem 6.2 applies. At this point it must be remarked that in the set of exceptional points on the real axis where the extended resolvents are not defined we should include the eigenvalues of $c_0^2(-\Delta_{i,D}+N)$. However we still have N at our disposal, and when we consider $\lambda\in K$, where K is any compact set contained on $(0,\infty)\setminus\sigma_+(A_e)$, we can take N so large in order that none of the eigenvalues of $-\Delta_{i,D}+N$ is contained in K. Moreover for $z\in\mathbf{C}^{\pm}$

$$(A_T-z)^{-1}=(-\Delta_{i,D}+N-z)^{-1}\oplus R_e(z), \tag{6.141}$$

where

$$R_e(z)=(A_e-z)^{-1}. \tag{6.142}$$

For $s\in\mathbf{R}$ we denote by $L_s^2(\Omega_e)$ the weighted space consisting of all complex valued measurable functions, $f(x,y)$, on Ω_e such that $\eta_s(x,y)$ $f(x,y)\in L^2(\Omega_e)$ with the natural norm

$$\left\|f\right\|_{L_s^2(\Omega_e)} = \left\|\eta_s\,f\right\|_{L^2(\Omega_e)}. \tag{6.143}$$

It follows from Theorem 6.2 that for any $\lambda \in K \subset (0, \infty) \setminus \sigma_+(A_e)$

$$R_e(\lambda \pm i\,0) = \lim_{\epsilon \downarrow 0} R_e(\lambda \pm i\,\epsilon) \tag{6.144}$$

exists on the uniform operator topology on $\mathcal{B}(L^2_s(\Omega_e),\ L^2_{-s}(\Omega_e))$, $s = \frac{1+\delta}{2}$, uniformly on $\lambda \in K$. Moreover, by Remark 6.3,

$$R_e(\lambda \pm i\,0) = -\chi_{\Omega_e}\, v(\lambda + M)^{-1}\,[I + (\lambda + M)^{-1}\, R^1_{\mp}\,((\lambda + M)^{-1})]v^{-1}\,\chi_{\Omega_e}, \tag{6.145}$$

where $\chi_{\Omega_e}(x, y)$ is the characteristic function of Ω_e. Moreover the functions

$$R_{e,\pm}(\lambda) = \begin{cases} R_e(\lambda), & Im\ \lambda \neq 0, \\ R_e(\lambda \pm i\,0), & \lambda \in (0, \infty) \setminus \sigma_+(A_e), \end{cases} \tag{6.146}$$

defined for $\lambda \in \mathbf{C}^{\pm} \cup ((0, \infty) \setminus \sigma_+(A_e))$with values on $\mathcal{B}(L^2_s(\Omega_e),\ L^2_{-s}(\Omega_e))$, $s = \frac{1+\delta}{2}$, are analytic for $\lambda \in \mathbf{C}^{\pm}$ and locally Hölder continuous for $\lambda \in (0, \infty) \setminus \sigma_+(A_e)$ with exponent γ as in Theorem 6.2

Note that by (6.145), Lemma 6.6 holds for $R_{e,\pm}(\lambda)$ if the spaces $L^2_{\pm s}(\mathbf{R}^{n+1})$ are replaced by $L^2_{\pm s}(\Omega_e)$ and if the neighborhood, O_c, of $c(x, y)$ is defined as in Definition 6.4, but for functions defined only on Ω_e and with the sup norm in (6.99) taken only on Ω_e, and with the sup norm in the right hand side of (6.117) taken only on Ω_e.

§7. The Generalized Fourier Maps and Acoustic Scattering Theory

In this section we will use the results on the limiting absorption principle for the perturbed acoustic propagator to construct the generalized Fourier maps for the perturbed propagator A_T and to study the acoustic scattering theory.

Let us first consider the unperturbed propagator A_0. By (2.10) for every Borel set $\Delta \subset (0, \infty)$, $\varphi,\ \psi \in \mathcal{H}_0$, $s > 1/2$,

$$(E_0(\Delta)\eta_{-s}\varphi,\ \eta_{-s}\psi)_{\mathcal{H}_0} = \int_{\Delta} (B(\lambda)\varphi,\ B(\lambda)\psi)_{\hat{\mathcal{H}}(\lambda)} d\lambda. \tag{7.1}$$

It follows (see Theorem 23, Chapter 11 of Royden 1968) that the Radon-Nikodym derivative of the spectral measure is given by

$$\frac{d}{d\lambda}(E_0(\lambda)\eta_{-s}\varphi,\ \eta_{-s}\psi)_{\mathcal{H}_0} = (B(\lambda)\varphi,\ B(\lambda)\psi)_{\hat{\mathcal{H}}(\lambda)}. \tag{7.2}$$

Moreover, by Stone's formula (see Kato 1976, Chapter 6, Section 5, Problem 5.7) and Theorem 2.4 for any $a, b > 0$, $\varphi,\ \psi \in \mathcal{H}_0$, and $s > 1/2$

$$(E_0(a,b)\eta_{-s}\varphi, \eta_{-s}\psi) = \lim_{\epsilon \downarrow 0} \frac{1}{2\pi i} \int_a^b \Big((R_0(\lambda + i\,\epsilon) -$$

$$- R_0(\lambda - i\,\epsilon))\eta_{-s}\varphi, \eta_{-s}\psi \Big)_{\mathcal{H}_0} =$$

$$= \int_a^b \frac{1}{2\pi i} \Big((R_0(\lambda + i\,0) - R_0(\lambda - i\,0))\ \eta_{-s}\varphi,\ \eta_{-s}\psi \Big)_{\mathcal{H}_0} d\lambda. \quad (7.3)$$

It follows that for $\lambda \geq 0$,

$$\frac{d}{d\lambda}(E_0(\lambda)\eta_{-s}\varphi,\ \eta_{-s}\psi)_{\mathcal{H}_0} = \Big(\frac{1}{2\pi i}(R_0(\lambda + i\,0) - R_0(\lambda - i\,0))\eta_{-s}\varphi,\ \eta_{-s}\psi \Big)_{\mathcal{H}_0}. \quad (7.4)$$

Let us denote by E_T the spectral family of A_T. Then by Stone's formula and Theorem 6.2 for any compact interval $[a,b] \subset \mathbf{R}^+ \setminus \sigma_+(A_T)$ $\varphi,\ \psi \in \mathcal{H}_0$, and $s = \frac{1+\delta}{2}$

$$(E_T(a,b)\eta_{-s}\varphi,\ \eta_{-s}\psi)_{\mathcal{H}} =$$

$$= \lim_{\epsilon \downarrow 0} \frac{1}{2\pi i} \int_a^b \Big((R_T(\lambda + i\,\epsilon) - R_T(\lambda - i\,\epsilon))\eta_{-s}\varphi,\ \eta_{-s}\psi \Big)_{\mathcal{H}} =$$

$$= \int_a^b \Big(\frac{1}{2\pi i}(R_T(\lambda + i\,0) - R_T(\lambda - i\,0))\eta_{-s}\varphi,\ \eta_{-s}\psi \Big)_{\mathcal{H}}. \quad (7.5)$$

It follows that for $\lambda \in (0,\infty) \setminus \sigma_+(A_T)$,

$$\frac{d}{d\lambda}(E_T(\lambda)\eta_{-s}\varphi,\ \eta_{-s}\psi)_{\mathcal{H}} = \Big(\frac{1}{2\pi i}(R_T(\lambda + i\,0) - R_T(\lambda - i\,0))\eta_{-s}\varphi,\ \eta_{-s}\psi \Big)_{\mathcal{H}}. \quad (7.6)$$

Moreover, by (6.98)

$$\Big(\frac{1}{2\pi i}(R_T(\lambda + i\,0) - R_T(\lambda - i\,0))\eta_{-s}\varphi,\ \eta_{-s}\psi \Big)_{\mathcal{H}} =$$

$$= \frac{-(\lambda + M)^{-2}}{2\pi i} \Big(\Big(R^1((\lambda + M)^{-1} - i0)\ -\ R^1((\lambda + M)^{-1} +$$

$$+ i0) \Big) v^{-1}\eta_{-s}\varphi, v^{-1}\eta_{-s}\psi \Big)_{\mathcal{H}_0} = \frac{-(\lambda + M)^{-2}}{2\pi i}$$

$$\lim_{\epsilon \downarrow 0}\left(\left(R^1((\lambda+i\epsilon+M)^{-1})-R^1((\lambda-i\epsilon+M)^{-1})\right)v^{-1}\eta_{-s}\varphi, v^{-1}\eta_{-s}\psi\right)_{\mathcal{H}_0}$$

$$= \frac{-(\lambda+M)^{-2}}{2\pi i}\left(\left(R^0((\lambda+M)^{-1}-i\,0)-R^0((\lambda+M)^{-1}+i\,0)\right)Q_{0,\mp}^{-1}\right.$$

$$\left.\cdot((\lambda+M)^{-1})\,v^{-1}\eta_{-s}\varphi,\ Q_{0,\mp}^{-1}((\lambda+M)^{-1})\,v^{-1}\,\eta_{-s}\psi\right)_{\mathcal{H}_0}, \qquad (7.7)$$

where we used (6.29) and the first resolvent equation (see Kato 1976, Chapter I, Problem 5.4, and page 173).

Moreover, as the reader can easily check for $z=\lambda+i\,\epsilon,\quad \epsilon\neq 0$

$$R_0(z)=-(\lambda+M)^{-1}\left(I+(\lambda+M)^{-1}\,R^0((z+M)^{-1})\right), \qquad (7.8)$$

and it follows by taking the limit when $\epsilon\downarrow 0$ that

$$(\lambda+M)^{-2}\left(R^0((\lambda+M)^{-1}-i\,0)-R^0((\lambda+M)^{-1}+i\,0)\right)$$

$$=R_0(\lambda+i\,0)-R_0(\lambda-i\,0). \qquad (7.9)$$

Then by (7.7)

$$\left(\frac{1}{2\pi i}(R_T(\lambda+i\,0)-R_T(\lambda-i\,0))\,\eta_{-s}\varphi,\ \eta_{-s}\psi\right)_{\mathcal{H}}$$

$$=\left(\frac{1}{2\pi i}(R_0(\lambda+i\,0)-R_0(\lambda-i\,0))\right.$$

$$\left.\cdot\,Q_{0,\mp}^{-1}((\lambda+M)^{-1})v^{-1}\eta_{-s}\varphi,\ Q_{0,\mp}^{-1}((\lambda+M)^{-1})v^{-1}\eta_{-s}\psi\right)_{\mathcal{H}_0}, \qquad (7.10)$$

and by (7.6)

$$\frac{d}{d\lambda}(E_T(\lambda)\eta_{-s}\varphi,\ \eta_{-s}\psi)_{\mathcal{H}}=\left(\frac{1}{2\pi i}(R_0(\lambda+i\,0)-R_0(\lambda-i\,0))\right.$$

$$\left.\cdot\,Q_{\mp}((\lambda+M)^{-1})v^{-1}\eta_{-s}\varphi,\ Q_{\mp}((\lambda+M)^{-1})v^{-1}\eta_{-s}\psi\right)_{\mathcal{H}_0}, \qquad (7.11)$$

where for $\sigma\in(0,\frac{1}{M})\setminus\sigma_+(H_1)$,

$$Q_{\pm}(\sigma)=(H_0-\sigma)R^1_{\pm}(\sigma)\equiv I-V\,R^1_{\pm}(\sigma). \qquad (7.12)$$

Note that

$$Q_{\pm}(\sigma)=Q_{0,\pm}^{-1}(\sigma). \qquad (7.13)$$

It follows from (7.2), (7.4) and (7.11) that for $\lambda \in (0,\infty) \setminus \sigma_+(A_T)$

$$\frac{d}{d\lambda}\Big(E_T(\lambda)\eta_{-s}\varphi,\, \eta_{-s}\psi\Big)_{\mathcal{H}} = \Big(B(\lambda)\eta_s\, Q_{\mp}((\lambda+M)^{-1})v^{-1}\eta_{-s}\varphi,$$

$$\cdot\, B(\lambda)\eta_s\, Q_{\mp}((\lambda+M)^{-1})v^{-1}\eta_{-s}\psi\Big)_{\hat{\mathcal{H}}(\lambda)}. \tag{7.14}$$

We denote by $\mathcal{H}_{ac}(A_T)$ the subspace of absolute continuity of A_T.

Let $F_T^{\pm}$ be the following operators from $\mathcal{H}_{ac}(A_T)$ onto $\hat{\mathcal{H}}$, defined first on vectors of the type

$$\psi = \sum_{j=1}^{N} E_T(I_j)\eta_{-s}\psi_j, \tag{7.15}$$

where the intervals $I_j \subset (0,\infty) \setminus \sigma_+(A_T)$, $\quad I_j \cap I_k = \emptyset,\ j \neq k, \quad \psi_j \in \mathcal{H}$, and $E_T(\,\cdot\,)$ denote the spectral family of A_T, as

$$F_T^{\pm}\psi = \sum_{j=1}^{N} \chi_{I_j}(\lambda)\, B(\lambda)\eta_s\, Q_{\mp}((\lambda+M)^{-1})v^{-1}\eta_{-s}\psi_j, \tag{7.16}$$

where $\chi_{I_j}(\lambda)$ denotes the characteristic function of the interval I_j.

By (7.14) and (7.16)

$$(F_T^{\pm}\psi,\, F_T^{\pm}\psi)_{\hat{\mathcal{H}}} = \sum_{j=1}^{N} \int_{I_j} d\lambda\, \frac{d}{d\lambda}\,(E_T(\lambda)\eta_{-s}\psi_j,\, \eta_{-s}\psi_j)_{\mathcal{H}} =$$

$$= \sum_{j=1}^{N}(E_T(I_j)\eta_{-s}\psi_j,\, E_T(I_j)\eta_{-s}\psi_j) = (\psi,\psi)_{\mathcal{H}}. \tag{7.17}$$

By (7.17) the operators $F_T^{\pm}$ are well defined, and are isometric from $\mathcal{H}_{ac}(A_T)$ into $\hat{\mathcal{H}}$.

Since the set of vectors of the form (7.15) is dense on $\mathcal{H}_{ac}(A_T)$, we extend the $F_T^{\pm}$ by continuity to operators defined on all of $\mathcal{H}_{ac}(A_T)$. We extend the $F_T^{\pm}$ as partially isometric operators on $\mathcal{H}$ by defining $F_T^{\pm} = 0$, on $\mathcal{H}_{pp}(A_T)$.

Then we have that

Lemma 7.1

Suppose that the local compactness assumption 3.2 is satisfied and that for some constants $C, \quad \delta > 0$

$$|c(x,y) - c_0(y)| \leq C\ (1 + |x| + |y|)^{-1-\delta}. \tag{7.18}$$

Then the generalized Fourier maps, $F_T^{\pm}$, defined above have the following properties

1. Kernel $(F_T^{\pm}) = \mathcal{H}_{pp}(A_T)$. (7.19)

2. The restrictions of $F_T^{\pm}$ to $\mathcal{H}_{ac}(A_T)$ are unitary operators from $\mathcal{H}_{ac}(A_T)$ onto $\hat{\mathcal{H}}$ and

$$A_T\ P_{ac}(A_T) = F_T^{\pm *}\ \lambda\ F_T^{\pm}, \tag{7.20}$$

where $P_{ac}(A_T)$ denotes the orthogonal projector onto $\mathcal{H}_{ac}(A_T)$, and λ denotes the operator of multiplication by the independent variable in the direct integral in (1.63).

Proof: (1) follows from the definition of the $F_T^{\pm}$. To prove that the restrictions of the $F_T^{\pm}$ to $\mathcal{H}_{ac}(A_T)$ are unitary onto $\hat{\mathcal{H}}$ it only remains to prove that the $F_T^{\pm}$ are onto.

Suppose that there is a $h(\lambda) \in \hat{\mathcal{H}}$ that is orthogonal to the range of the $F_T^{\pm}$. For $j = 0, 1, 2, 3, \cdots$ we denote by $I_j = (\lambda_j, \lambda_{j+1}) \setminus \sigma_+(A_T)$, where $\lambda_0 = 0$. Then for all Borel sets I', with $\overline{I}' \subset I_j$, and all $\psi \in \mathcal{H}$,

$$\left(F_T^{\pm}\ E_T(I')\eta_{-s}\psi,\ h\right)_{\hat{\mathcal{H}}} =$$

$$= \int_{I'} \left(B(\lambda)\eta_s\ Q_{\mp}((\lambda + M)^{-1})v^{-1}\eta_{-s}\psi,\ h\right)_{\hat{\mathcal{H}}(\lambda_j)} d\lambda, \tag{7.21}$$

where we used (7.16). Note that for $\lambda \in (\lambda_j, \lambda_{j+1})$, $\hat{\mathcal{H}}(\lambda) = \hat{\mathcal{H}}(\lambda_j)$.

By (7.13) the operators $Q_{\pm}((\lambda+M)^{-1})$, $\lambda \in I'$ are onto $L^2_s(\mathbf{R}^{n+1})$,$s = \frac{1+\delta}{2}$. Then since F is unitary it follows from (2.10) that the set of all linear combinations of all functions of the form

$$\chi_{I'}(\lambda)\ B(\lambda)\eta_s\ Q_{\mp}((\lambda + M)^{-1})v^{-1}\eta_{-s}\psi, \tag{7.22}$$

where $\overline{I}' \subset I_j$, $\psi \in \mathcal{H}$, is dense on $L^2(I_j,\ \hat{\mathcal{H}}(\lambda_j))$. Then by Lemmas 4.4 and 4.6 of Kuroda 1973, $h(\lambda) = 0$ for a.e. $\lambda \in (\lambda_j,\ \lambda_{j+1})$, and since this is true for all $j = 0, 1, 2, 3, \cdots$, we have that $h(\lambda) = 0$ for a.e. $\lambda \in (0, \infty)$, and in consequence the $F_T^{\pm}$ are onto $\hat{\mathcal{H}}$.

Finally it follows from (7.17) that for all ψ's of the form (7.16),

$$F_T^{\pm} A_T \psi = \lambda(F_T^{\pm} \psi)(\lambda), \tag{7.23}$$

and since that set of ψ's is dense on $\mathcal{H}_{ac}(A_T)$, for all $\psi \in D(A_T)$

$$F_T^{\pm} A_T P_{ac}(A_T) \psi = \lambda(F_T^{\pm} \psi)(\lambda). \tag{7.24}$$

Then

$$F_T^{\pm} A_T P_{ac}(A_T) = \lambda F_T^{\pm}, \tag{7.25}$$

and (7.20) follows since $F_T^{\pm}$ are unitary from $\mathcal{H}_{ac}(A_T)$ onto $\hat{\mathcal{H}}$.

Q. E. D.

Let J be the operator of identification from $\mathcal{H}_0$ onto $\mathcal{H}$ given by

$$(J \varphi)(x, y) = \varphi(x, y). \tag{7.26}$$

J is clearly a bijection from $\mathcal{H}_0$ onto $\mathcal{H}$.

The wave operators for the transmission problem are defined as

$$W_T^{\pm} = s - \lim_{t \to \pm \infty} e^{itA_T} J e^{-itA_0}, \tag{7.27}$$

provided that the strong limits exist.

The scattering operator is defined as

$$S_T = W_T^{+*} W_T^{-}. \tag{7.28}$$

The relation between the wave and scattering operators (7.27) and (7.28) and those for the equations (1.1) and (3.17) is well known. See Kato 1967, and Reed and Simon 1979, Section 10 of Chapter XI.

We will prove that the wave operators satisfy the invariance principle. For this purpose we introduce the following class of functions.

Definition 7.2

We say that a real valued measurable function, $f(\lambda)$, defined on $(0, \infty)$ is admissible if for all $\varphi(\lambda) \in L^2(0, \infty)$,

$$\lim_{t \to \infty} \int_0^{\infty} \left| \int_0^{\infty} e^{-itf(\lambda) - is\lambda} \varphi(\lambda) d\lambda \right|^2 ds = 0. \tag{7.29}$$

Note that in particular $f(\lambda) = \lambda$ is admissible. More generally it is proven in Kato 1976, Lemma 4.6, Section 4 of Chapter 10, that $f(\lambda)$ is admissible if it satisfies the following conditions: the axis $(0, \infty)$ can be divided into a finite number of subintervals in such a way that in each open subinterval, $f(\lambda)$ is differentiable with $\frac{d}{d\lambda} f(\lambda)$ continuous, locally of bounded variation, and positive.

We have that

Lemma 7.3

Under the conditions of Lemma 7.1 the wave operators $W_T^{\pm}$ exist and are unitary from $\mathcal{H}_0$ onto $\mathcal{H}_{ac}(A_T)$. Moreover the scattering operator, S_T, is unitary. The following stationary formulas for the wave operators hold

$$W_T^{\pm} = F_T^{\pm *}\, F. \tag{7.30}$$

The invariance principle is satisfied: for any admissible function $f(\lambda)$,

$$W_T^{\pm} = s - \lim_{t \to \pm\, \infty} e^{itf(A_T)}\, J\, e^{-itf(A_0)}, \tag{7.31}$$

where $f(A_T)$ and $f(A_0)$ are defined by functional calculus of selfadjoint operators.

Proof: Once Theorem 6.2 and Lemma 7.1 are proved, the results on the wave and scattering operators stated in the Lemma are classical. We briefly indicate the proof here following the argument given in Theorem 3, Section 6, Chapter 5, of Kuroda 1980.

We define the wave operators as

$$W_T^{\pm} = F_T^{\pm *}\, F, \tag{7.32}$$

where F is as in Theorem 1.1. Then the $W_T^{\pm}$ are unitary from $\mathcal{H}_0$ onto $\mathcal{H}_{ac}(A_T)$ by Theorem 1.1 and Lemma 7.1, and it follows that S_T is unitary. We prove below that with the $W_T^{\pm}$ defined as in (7.32), formulas (7.31) hold. Then the existence of the wave operators defined as in (7.27) follows since $f(\lambda) = \lambda$ is admissible.

We consider the case of W_T^{+}, the one of W_T^{-} being similar. First note that since $F_T^{\pm *}\, F$ is unitary from $\mathcal{H}_0$ onto $\mathcal{H}_{ac}(A_T)$ and for all $\varphi \in \mathcal{H}_0$,

$$\lim_{t \to \infty} \|e^{itf(A_T)}\, J\, e^{-itf(A_0)} \varphi\|_{\mathcal{H}} = \|\varphi\|_{\mathcal{H}_0}, \tag{7.33}$$

by the Rellich local compactness theorem, it is enough to prove that

$$\lim_{t\to\infty}\left(e^{-itf(A_T)}E_T(\Delta_1)\eta_{-s}\varphi,\ J\,e^{-itf(A_0)}E_0(\Delta_0)\eta_{-s}\psi\right)_{\mathcal{H}}=$$

$$=\left(E_T(\Delta_1)\eta_{-s}\varphi,\ F_T^{+*}\ F_0E_0(\Delta_0)\eta_{-s}\psi\right)_{\hat{\mathcal{H}}},\tag{7.34}$$

for all bounded and closed sets $\Delta_0\subset(0,\infty)$ and $\Delta_1\subset(0,\infty)\setminus\sigma_+(A_T)$, and all $\varphi\in\mathcal{H}$, $\psi\in\mathcal{H}_0$.

Moreover, since

$$|1-v(x,y)|\leq C\ (1+|x|+|y|)^{-1-\delta},\tag{7.35}$$

it also follows from the Rellich local compactness theorem that

$$\lim_{t\to+\infty}\left\|J(v-1)\ e^{-itf(A_0)}\ E_0(\Delta)\eta_{-s}\psi\right\|_{\mathcal{H}}=0.\tag{7.36}$$

Then (7.34) is equivalent to

$$\lim_{t\to+\infty}\ (e^{-itf(A_T)}\ E_T(\Delta_1)\eta_{-s}\varphi,\ J\ v\ e^{-itf(A_0)}\ E_0(\Delta_0)\eta_{-s}\psi)_{\mathcal{H}}\ =$$

$$=\ (F_T^{\pm}\ E_T(\Delta)\eta_{-s}\varphi,\ F_0\ E_0(\Delta_0)\eta_{-s}\psi)_{\hat{\mathcal{H}}}.\tag{7.37}$$

It follows from (6.26), (6.97), and (7.8), that for $\lambda\in(0,\infty)$, $\epsilon>0$

$$\frac{1}{2\pi\ i}\ (R_T(\lambda+i\ \epsilon)-R_T(\lambda-i\ \epsilon))=\frac{v}{2\pi\ i}\ (R_0(\lambda+i\ \epsilon)-R_0(\lambda-i\ \epsilon))$$

$$\cdot Q((\lambda+i\ \epsilon+M)^{-1})\ v^{-1}-\frac{v}{2\pi\ i}\ (\lambda-i\ \epsilon+M)^{-2}R^0((\lambda-i\ \epsilon+M)^{-1})$$

$$\cdot\left(Q((\lambda+i\ \epsilon+M)^{-1})-Q((\lambda-i\ \epsilon+M)^{-1})\right)\ v^{-1}+$$

$$+\ \frac{\epsilon}{\pi((\lambda+M)^2+\epsilon^2)}-\frac{\epsilon}{\pi((\lambda+M)^2+\epsilon^2)}v\ Q((\lambda+i\ \epsilon+M)^{-1})v^{-1}.\tag{7.38}$$

Then it follows from (2.10), (7.2), (7.4), (7.6), (7.16) and (7.38) that

$$(e^{-itf(A_T)}\ E_T(\Delta_1)\eta_{-s}\varphi,\ J\ v\ e^{-itf(A_0)}\ E_0(\Delta_0)\eta_{-s}\psi)_{\mathcal{H}}\ =$$

$$=\ (F^+\ E_T(\Delta_1)\eta_{-s}\varphi,\ F_0\ E_0(\Delta_0)\psi)_{\hat{\mathcal{H}}}\ +$$

$$+\ \int_{\Delta} e^{-itf(\lambda)}\ \lim_{\epsilon\downarrow 0}\ h_{\epsilon,t}(\lambda)d\lambda,\tag{7.39}$$

where

$$h_{\epsilon,t}(\lambda) = -\Bigg(\frac{1}{2\pi\, i}(\lambda - i\,\epsilon + M)^{-2}\, R^0((\lambda - i\,\epsilon + M)^{-1})$$

$$\cdot\ \Big(Q((\lambda + i\,\epsilon + M)^{-1}) - Q((\lambda - i\,\epsilon + M)^{-1})\Big)$$

$$\cdot\ v^{-1}\,\eta_{-s}\varphi,\ e^{-itf(A_0)}\, E_0(\Delta_0)\eta_{-s}\psi\Bigg)_{\mathcal{H}_0}. \tag{7.40}$$

But

$$h_{\epsilon,t}(\lambda) = (g_\epsilon(\lambda),\quad J_{\epsilon,t}(\lambda))_{\mathcal{H}_0}, \tag{7.41}$$

where

$$g_\epsilon(\lambda) = -\,(2\pi\, i(\lambda - i\,\epsilon + M))^{-1}\,\eta_s\Big(Q((\lambda + i\,\epsilon + M)^{-1}) -$$

$$-\,Q((\lambda - i\,\epsilon + M)^{-1})\Big)v^{-1}\eta_{-s}\varphi, \tag{7.42}$$

$$J_{\epsilon,t}(\lambda) = \int_{\Delta_0} e^{-itf(\rho)}\,\frac{B^*(\rho)B(\rho)}{\lambda - \rho + i\,\epsilon}\,\psi\, d\rho. \tag{7.43}$$

Moreover,

$$\lim_{\epsilon\downarrow 0} g_\epsilon(\lambda) = g_+(\lambda) =$$

$$= (2\pi\, i(\lambda + M))^{-1}\,\eta_s\Big(Q_+((\lambda + M)^{-1}) - Q_-((\lambda + M)^{-1})\Big)\, v^{-1}\eta_{-s}\varphi, \tag{7.44}$$

in the strong topology on $\mathcal{H}_0$ uniformly for $\lambda \in \Delta_1$. Furthermore,

$$J_{\epsilon,t}(\lambda) = \ell_\epsilon\ *\ k_t, \tag{7.45}$$

where

$$\ell_\epsilon(\lambda) = (\lambda + i\,\epsilon)^{-1}, \tag{7.46}$$

$$k_t(\rho) = \chi_{\Delta_0}(\rho)\, e^{-itf(\rho)}\, B^*(\rho)B(\rho)\psi, \tag{7.47}$$

where $\chi_{\Delta_0}(\rho)$ denotes the characteristic function of Δ_0.

Then denoting respectively by $\hat{}$ and $\check{}$ the Fourier and inverse Fourier transforms on $L^2(\mathbf{R}, \mathcal{H}_0)$, we obtain by the convolution theorem that

$$\hat{J}_{\epsilon,t}(p) = -2\pi\, i\, e^{-\epsilon p}\chi_{(0,\infty)}(p)\, \hat{k}_t(p), \tag{7.48}$$

where $\chi_{(o,\infty)}(p)$ is the characteristic function of $(0, \infty)$. Then

$$\lim_{\epsilon \downarrow 0} \hat{J}_{\epsilon,t}(p) = \hat{J}_{+,t}(p) \equiv -2\pi\, i\, \chi_{(0,\infty)}(p)\, \hat{k}_t(p), \tag{7.49}$$

strongly on $L^2(\mathbf{R}, \mathcal{H}_0)$, and by taking the inverse Fourier transform in $L^2(\mathbf{R}, \mathcal{H}_0)$,

$$\lim_{\epsilon \downarrow 0} J_{\epsilon,t}(\lambda) = J_{+,t}(\lambda) \equiv (-2\pi\, i\, \chi_{(0,\infty)}(p)\, \hat{k}_t(p))\check{}(\lambda), \tag{7.50}$$

in the strong topology on $L^2(\mathbf{R}, \mathcal{H}_0)$. It follows that

$$I(t) \equiv \int_\Delta e^{-itf(\lambda)} \lim_{\epsilon \downarrow 0} h_{\epsilon,t}(\lambda) d\lambda =$$

$$= \int_\Delta e^{-itf(\lambda)} (g_+(\lambda),\ J_{+,t}(\lambda))_{\mathcal{H}_0} d\lambda. \tag{7.51}$$

Moreover

$$|I(t)| \leq C \left\| J_{+,t}(\lambda) \right\|_{L^2(\mathbf{R},\mathcal{H}_0)} = C \left\| \hat{J}_{+,t}(p) \right\|_{L^2(\mathbf{R},\mathcal{H}_0)} \leq$$

$$\leq C \left(\int_0^\infty dp \left\| \int_0^\infty e^{-itf(\rho)-ip\rho} \chi_{\Delta_0}(\rho)\, B^*(\rho) B(\rho) \psi \right\|^2_{\mathcal{H}_0} \right)^{1/2} \tag{7.52}$$

It follows by (7.29) and (7.52) that

$$\lim_{t \to \infty} I(t) = 0, \tag{7.53}$$

and (7.37) follows from (7.39), (7.51), and (7.53)

Q. E. D.

Let us denote

$$\hat{S}_T = F\ S\ F^* = F_T^+\ F_T^{-*}. \tag{7.54}$$

In the Lemma below we obtain a representation of the scattering operator $\hat{S}_T$ in terms of the "scattering matrix".

Lemma 7.4

Suppose that the conditions of Lemma 7.1 are satisfied. For $\lambda \in (0, \infty) \setminus \sigma_+(A_T)$ and $s = \frac{1+\delta}{2}$ denote

$$S_T(\lambda) = I + 2\,\pi\, i(\lambda + M)^2 B(\lambda) \eta_s\, Q_-((\lambda + M)^{-1}) V\, \eta_s\, B^*(\lambda). \tag{7.55}$$

Then $S_T(\lambda)$ is an unitary operator on $\hat{\mathcal{H}}(\lambda)$ with inverse given by

$$S_T^{-1}(\lambda) = I - 2\,\pi\, i(\lambda + M)^2 B(\lambda)\eta_s\, Q_+((\lambda + M)^{-1})V\ \eta_s\ B^*(\lambda). \quad (7.56)$$

Moreover $\hat{S}_T$ and $\hat{S}_T^{-1}$ are respectively the direct integral of $S_T(\lambda)$ an $S_T^{-1}(\lambda)$

$$\hat{S}_T = \oplus \int_0^\infty S_T(\lambda) d\lambda, \quad (7.57)$$

$$\hat{S}_T^{-1} = \oplus \int_0^\infty S_T^{-1}(\lambda) d\lambda. \quad (7.58)$$

That is to say for any

$$\varphi(\lambda) \in \hat{\mathcal{H}} = \oplus \int_0^\infty \hat{\mathcal{H}}(\lambda) d\lambda, \quad (7.59)$$

$$(\hat{S}_T \varphi)(\lambda) = \hat{S}_T(\lambda)\varphi(\lambda), \qquad \text{a.e. } \lambda, \quad (7.60)$$

$$(\hat{S}_T^{-1} \varphi)(\lambda) = \hat{S}_T^{-1}(\lambda)\varphi(\lambda), \qquad \text{a.e. } \lambda. \quad (7.61)$$

Moreover $S_T(\lambda) - I$ is a compact operator, and the function $\lambda \to S_T(\lambda)$ from $(0, \infty) \setminus \sigma_+(A_T)$ into $\mathcal{B}(\hat{\mathcal{H}}(\infty))$ is locally Hölder continuous with exponent γ that satisfies $\gamma < 1$, $\gamma < (s - 1/2)$, if $\lambda \neq \lambda_j$, for some $j = 1, 2, 3 \cdots$, and $\gamma < 1/2$, $\gamma < (s - 1/2)$, if $\lambda = \lambda_j$, for some $j = 1, 2, 3, \cdots$.

Proof: This is a classical result. We follow the proof of Theorem 2, Section 7, Chapter 5 of Kuroda 1980. Equation (7.60) is equivalent to

$$(F_T^+ \varphi)(\lambda) = S_T(\lambda)(F_T^- \varphi)(\lambda), \quad (7.62)$$

for all $\varphi \in \mathcal{H}_{ac}(A_T)$.

By continuity it is enough to prove that (7.62) holds for all φ of the form

$$\varphi = E_T(I)\eta_{-s}\psi, \quad (7.63)$$

with I any closed and bounded set, $I \subset (0, \infty) \setminus \sigma_+(A_T)$. First note that by (6.28), (7.2), (7.4), and (7.8)

$$Q_{0,+}((\lambda + M)^{-1}) - Q_{0,-}((\lambda + M)^{-1}) = 2\,\pi\, i(\lambda + M)^2\ V\eta_s\ B^*(\lambda)B(\lambda)\eta_s. \quad (7.64)$$

Denote

$$f_\pm(\lambda) = F_T^\pm E_T(I)\eta_{-s}\psi = \chi_I(\lambda)B(\lambda)\eta_s\ Q_\mp((\lambda + M)^{-1})v^{-1}\eta_{-s}\psi. \quad (7.65)$$

Then

$$\begin{aligned} f_+(\lambda) &= f_-(\lambda) + \chi_I(\lambda)B(\lambda)\eta_s(Q_- - Q_+)v^{-1}\eta_{-s}\psi = \\ &= f_-(\lambda) + \chi_I(\lambda)B(\lambda)\eta_s\, Q_-(Q_{0,+} - Q_{0,-})Q_+\, v^{-1}\eta_{-s}\psi = \\ &= S_T(\lambda)f_-(\lambda), \end{aligned} \tag{7.66}$$

where we used (7.64) and $Q_{0,\pm} = Q_\pm^{-1}$.

Define $S'_T(\lambda)$ by the right hand side of (7.56), then as above we prove that

$$(S_T^{-1}\varphi)(\lambda) = S'_T(\lambda)\varphi(\lambda), \tag{7.67}$$

and it follows by taking $\varphi(\lambda)$ constant in small intervals that

$$S'_T(\lambda) = S_T^{-1}(\lambda). \tag{7.68}$$

The fact that $S_T(\lambda)$ is unitary follows similarly since $\hat{S}_T$ is unitary.

That $S_T(\lambda) - I$ is compact is immediate from Lemma 4.5 and (7.55). The local Hölder continuity of $S_T(\lambda)$ in λ is also obvious from (7.55).

Q. E. D.

In the following remarks we study the continuity of the wave and scattering operators on the perturbation.

Remark 7.5

Let $\left\{c_m\right\}_{m=1}^{\infty}$ be real valued measurable functions that satisfy (3.2) and (7.18) and such that

$$\lim_{m\to\infty} \|(1+|x|+|y|)^{1+\delta}\,(c_m - c)\|_\infty = 0. \tag{7.69}$$

We denote by $W_{T,m}^{\pm}$ and $S_{T,m}$ the wave and scattering operators corresponding to the operator $A_{T,m} = -c_m^2\Delta_T$. Define

$$\tilde{W}_{T,m}^{\pm} = L_m W_{T,m}^{\pm}, \tag{7.70}$$

where L_m is defined as in (4.32) with $v_m = c_m c_0^{-1}$ instead of v. Note that

$$W_{\pm,m}^{\sim} = s - \lim_{t\to\pm\,\infty} e^{it\tilde{A}_{T,m}}\, e^{-itA_0}, \tag{7.71}$$

where $\tilde{A}_{T,m} = L\, A_{T,m} L^{-1}$. Similarly denote

$$\tilde{W}_T^{\pm} = L\, W_T^{\pm} = s - \lim_{t \to \pm\, \infty} e^{it\tilde{A}_T}\, e^{-itA_0}. \tag{7.72}$$

Then

$$s - \lim_{m \to \infty} \tilde{W}_{T,m}^{\pm} = \tilde{W}_T^{\pm}, \tag{7.73}$$

$$s - \lim_{m \to \infty} S_{T,m} = S_T, \tag{7.74}$$

in the strong operator topology on $\mathcal{H}_0$.

Proof: (7.73) follows from Lemma 6.5, Lemma 6.6, (7.16), and (7.30). Note that $\tilde{W}_{T,m}^{\pm}$ and $\tilde{W}_T^{\pm}$ are isometric. Moreover, since $S_{T,m}$ and S_T are unitary, (7.74) follows from (7.28) and (7.73).

Remark 7.6

We denote by $S_{T,c}(\lambda)$ the operator (7.55) for $A_{T,c}$. Then for any compact set of $K \subset (0,\infty) \setminus \sigma_+(A_{T,c})$ there is a neighborhood, O_c, of $c(x,y)$ (see Definition 6.4) such that for $d(x,y) \in O_c$, $\quad K \subset (0,\infty) \setminus \sigma_+(A_{T,d})$ and

$$\left\| S_{T,d}(\lambda) - S_{T,c}(\lambda) \right\|_{\hat{\mathcal{H}}(\infty)} \leq C \left\| (1 + |x| + |y|)^{1+\delta}(c-d) \right\|_{\infty}, \tag{7.75}$$

for some constant C, and all $\lambda \in K$.

Proof: The result is immediate from Lemma 6.5, Lemma 6.6, and (7.55)

Remark 7.7

For any $f(\omega) \in L^2(S_c)$ we denote

$$\phi_{0,f}(x,y,\lambda) = \int_{S_c} \phi_0(x,y,\lambda,\omega) f(\omega) d\omega, \tag{7.76}$$

where $\phi_0(x,y,\lambda,\omega)$ is defined in (1.44) to (1.46).

Moreover for any $f(\nu) \in L^2(S_1^{n-1})$ we define

$$\phi_{j,f}(x,y,\lambda) = \int_{S_1^{n-1}} \phi_j(x,y,\lambda,\nu) f(\nu) d\nu, \tag{7.77}$$

for $j = 1,2,3\cdots$, with $\phi_j(x,y,\lambda,\nu)$ defined on (1.49).

Note that by (2.7) for any $f = \bigoplus_{j=0}^{\infty} f_j \in \hat{\mathcal{H}}$

$$B^*(\lambda) f = \eta_{-s} \sum_{j=0}^{N} \phi_{j,f_j}(x,y,\lambda), \tag{7.78}$$

where N is such that $\lambda_N < \lambda \leq \lambda_{N+1}$.

Denote

$$T_T(\lambda) = S_T(\lambda) - I. \tag{7.79}$$

Then by (7.55) and (7.78)

$$\left(T_T(\lambda) f, g\right)_{\hat{\mathcal{H}}(\lambda)} = 2\,\pi\, i(\lambda + M)^2 \sum_{j,k=0}^{N} \left(V\, \phi_{j,f_j}^{-,T},\, \phi_{k,g_k}\right)_{\mathcal{H}_0}, \tag{7.80}$$

where

$$\phi_{j,f_j}^{-,T} = \phi_{j,f_j} - R_-^1((\lambda+M)^{-1})\, V\, \phi_{j,f_j}. \tag{7.81}$$

Suppose moreover that for some $\delta > 0$,

$$|c(x,y) - c_0(y)| \leq C\, (1 + |x| + |y|)^{-n-1-\delta}. \tag{7.82}$$

Then since ϕ_j, $\quad j = 0,1,2,3,\cdots$, are bounded functions (see Appendix 1)

$$\phi_0(x,y,\lambda,\omega),\ \phi_j(x,y,\lambda,\nu) \in L^2_{-s}(\mathbf{R}^{n+1}), \tag{7.83}$$

with $s = \frac{n+1+\delta}{2}$, and we can define

$$\phi_0^{-,T}(x,y,\lambda,\omega) = \phi_0(x,y,\lambda,\omega) - R_-^1((\lambda+M)^{-1}) V \phi_0, \tag{7.84}$$

and for $j = 1,2,3,\cdots$

$$\phi_j^{-,T}(x,y,\lambda,\nu) = \phi_j(x,y,\lambda,\nu) - R_-^1((\lambda+M)^{-1}) V \phi_j, \tag{7.85}$$

for $\lambda \in (0,\infty) \setminus \sigma_+(A_T)$.

In this case we can invert the order of integration on (7.80), and $T_T(\lambda)$ is an integral operator with kernel

$$(T_T(\lambda) f)_k = 2\,\pi\, i(\lambda+M)^2 \sum_{j=0}^{N} \int T_{T,k,j}(\lambda, \nu_k, \acute{\nu}_j) f_j(\acute{\nu}_j) d\,\acute{\nu}_j, \tag{7.86}$$

where

$$T_{T,k,j}(\lambda, \nu_k, \acute{\nu}_j) = (V\phi_j^{-,T}(x,y,\lambda,\acute{\nu}_j),\ \phi_k(x,y,\lambda,\nu_k))_{\mathcal{H}_0}, \tag{7.87}$$

where $\nu_0,\ \acute{\nu}_0 \in S_c$, and $\nu_j,\ \acute{\nu}_j \in S_1^{n-1},\quad j = 1,2,3.$

Moreover, it follows from (7.87) that $T_T(\lambda)$ is a Hilbert-Schmidt operator. Let us see that the $v\ \phi_j^{-,T}$ are generalized eigenfunctions of A_T.

Since (see Appendix 1)

$$(A_0 - \lambda)\phi_j(x,y,\lambda,\nu) = 0, \tag{7.88}$$

where for $j = 0,\quad \nu \in S_c$, and for $j = 1,2,3,\cdots,\ \nu \in S_1^{n-1}$, and since by Lemma 4.1 (by taking adjoints) $(A_0 + M)^{-1}$ is bounded from $L^2_{-s}(\mathbf{R}^{n+1})$ into $L^2_{-s}(\mathbf{R}^{n+1}),\quad s \geq 0$, we have that

$$(H_0 - (\lambda+M)^{-1})\phi_j = -(\lambda+M)^{-1}(A_0+M)^{-1}(A_0-\lambda)\phi_j = 0, \tag{7.89}$$

for $j = 0,1,2,3,\cdots$.

Then by (7.84), (7.85), and Lemma 4.5

$$(H_1 - (\lambda+M)^{-1})\ \phi_j^{-,T} = V\ \phi_j - V\ \phi_j = 0, \tag{7.90}$$

But then

$$(H_0 - (\lambda+M)^{-1})\phi_j^{-,T} = -V\ \phi_j^{-,T}, \tag{7.91}$$

and it follows as in the proof of (6.65) that

$$\tilde{A}_T\ \phi_j^{-,T} = \lambda\ \phi_j^{-,T}, \tag{7.92}$$

in distribution sense in Ω, for $j = 0,1,2,\cdots$, and by (4.34) also

$$A_T\ v\ \phi_j^{-,T} = \lambda\ v\ \phi_j^{-,T}, \tag{7.93}$$

in distribution sense in Ω.

Remark 7.8

As in Remark 6.7 in the particular case when $\overline{\Omega}_i = \phi$, we obtain the perturbed acoustic propagator, A, in all space $\mathbf{R}^{n+1}$.

Let us denote by $E(\lambda)$ the family of spectral projectors of A.

It follows from (6.120), (7.6), and the first resolvent equation (see Kato 1976, Chapter I, Problem 5.4 and page 173) that for $\lambda > 0,\quad \varphi,\ \psi \in \mathcal{H}$, $s = \frac{1+\delta}{2}$,

$$\frac{d}{d\lambda}\left(E(\lambda)\eta_{-s}\varphi,\eta_{-s}\psi\right)_{\mathcal{H}} =$$

$$= \lim_{\epsilon\downarrow 0} \frac{1}{2\pi i}\left((R(\lambda+i\epsilon)-R(\lambda-i\epsilon))\eta_{-s}\varphi,\ \eta_{-s}\psi\right)_{\mathcal{H}} =$$

$$= \lim_{\epsilon\downarrow 0} \frac{1}{2\pi i}\left[\left((R_0(\lambda+i\epsilon)-R_0(\lambda-i\epsilon))\ G(\lambda\pm i\epsilon)v^{-2}\eta_{-s}\varphi,\right.\right.$$

$$\left.\cdot\, G(\lambda\pm i\,\epsilon)v^{-2}\eta_{-s}\psi\right)_{\mathcal{H}_0} + 2i\epsilon\left(\left(\frac{c_0^2}{c^2}-1\right)R_0(\lambda\pm\ i\ \epsilon)\ G(\lambda\pm i\epsilon)v^{-2}\eta_{-s}\varphi,\right.$$

$$\left.\left.R_0(\lambda\pm i\epsilon)\ G(\lambda\pm i\epsilon)v^{-2}\eta_{-s}\psi\right)_{\mathcal{H}_0}\right] =$$

$$= \left(B(\lambda)\eta_s G_\pm(\lambda)v^{-2}\eta_{-s}\varphi,\ B(\lambda)\eta_s G_\pm(\lambda)v^{-2}\eta_{-s}\psi\right)_{\mathcal{H}_0}, \tag{7.94}$$

where we used (7.2), (7.4), and Theorem 2.4, and where

$$G_\pm(\lambda) \equiv \lim_{\epsilon\downarrow 0} G(\lambda\ \pm\ i\ \epsilon) = I + \lambda(v^{-2}-I)R_\pm(\lambda)v^2. \tag{7.95}$$

Then

$$\frac{d}{d\lambda}\left(E(\lambda)\eta_{-s}\varphi,\ \eta_{-s}\psi\right)_{\mathcal{H}} = \left(D^\pm(\lambda)\eta_{-s}\varphi,\ D^\pm(\lambda)\eta_{-s}\psi\right)_{\hat{\mathcal{H}}(\lambda)}, \tag{7.96}$$

where

$$D^\pm(\lambda) = B(\lambda)\eta_s G_\pm(\lambda)v^{-2}\ =\ B(\lambda)\eta_s(I+\lambda(v^{-2}-1)R_\pm v^2)v^{-2}. \tag{7.97}$$

Then for ψ as in (7.15)

$$F^\pm\psi = \sum_{j=1}^{N} \chi_{I_j}(\lambda)D^\pm(\lambda)\eta_{-s}\psi_j. \tag{7.98}$$

Moreover since by (6.125), (7.2), and (7.4)

$$G_{0,+}(\lambda) - G_{0,-}(\lambda) = 2\pi i\lambda(1-v^{-2})\eta_s\ B^*(\lambda)B(\lambda)\eta_s, \tag{7.99}$$

it follows as in the proof of Lemma 7.4 that the operators (7.55) and (7.56) have the following representation (we use below the notation $S(\lambda)$, $S^{-1}(\lambda)$ instead of $S_T(\lambda)$, $S_T^{-1}(\lambda)$ to emphasize that we consider now the case of all of $\mathbf{R}^{n+1}$)

$$S(\lambda) = I - 2\pi i\lambda\ B(\lambda)\eta_s\ G_+(\lambda)(1-v^{-2})\eta_s\ B^*(\lambda), \tag{7.100}$$

$$S^{-1}(\lambda) = I + 2\pi i\lambda\ B(\lambda)\eta_s\ G_-(\lambda)(1 - v^{-2})\eta_s\ B^*(\lambda). \tag{7.101}$$

Moreover, denoting

$$T(\lambda) = S(\lambda) - I, \tag{7.102}$$

it follows from (7.78) and (7.100) that for $f,\ g \in \hat{\mathcal{H}}$

$$\Big(T(\lambda)f,\ g\Big)_{\hat{\mathcal{H}}(\lambda)} = -2\pi i\lambda \sum_{j,k=0}^{N} \Big((1 - v^{-2})\phi^-_{j,f_j},\quad \phi^0_{k,g_k}\Big)_{\mathcal{H}_0}, \tag{7.103}$$

where

$$\phi^-_{j,f_j}(x, y, \lambda) = \phi^0_{j,f_j} + \lambda\ R_+(1 - v^2)\phi^0_{j,f_j}. \tag{7.104}$$

Moreover, if (7.82) is satisfied, $T(\lambda)$ is an integral operator with kernel

$$(T(\lambda)f)_k = -2\pi i\lambda \sum_{j=0}^{N} \int T_{k,j}(\lambda, \nu_k, \acute{\nu}_j) f_j(\acute{\nu}_j) d\,\acute{\nu}_j, \tag{7.105}$$

where

$$T_{k,j}(\lambda, \nu_k, \acute{\nu}_j) = \Big((1 - v^{2})\ \phi_j\,(x, y, \lambda, \acute{\nu}_j),\ \phi_k(x, y, \lambda, \nu_k)\Big)_{\mathcal{H}_0}, \tag{7.106}$$

and

$$\phi^-_j(x, y, \lambda, \nu_j) = \phi_j(x, y, \lambda, \nu_j) + \lambda\ R_+(\lambda)(1 - v^2)\phi_j, \tag{7.107}$$

where $\nu_0 \in S_c$, and $\nu_j \in S_1^{n-1}$, $\quad j = 1, 2, 3\cdots$.

Note that

$$(A - \lambda)\phi^-_j(x, y, \lambda, \nu_j) = \lambda(v^2 - 1)\phi_j + \lambda(1 - v^2)\phi_j = 0. \tag{7.108}$$

Then ϕ^-_j is a generalized eigenfunction of A:

$$A\ \phi^-_j(x, y, \lambda, \nu_j) = \lambda\ \phi^-_j(x, y, \lambda, \nu_j), \tag{7.109}$$

$j = 0, 1, 2, \cdots$.

Remark 7.9

As in Remark 6.8 the case of an exterior domain Ω_e is a particular case of the transmission problem. Let J_e be the following bounded operator from $\mathcal{H}_0$ onto $\mathcal{H}_e$:

$$(J_e\varphi)(x,y) = \chi_{\Omega_e}(x,y)\varphi(x,y), \tag{7.110}$$

where $\chi_{\Omega_e}(x,y)$ is the characteristic function of Ω_e.

The wave operators are defined as

$$W_e^{\pm} = s - \lim_{t\to\pm\infty} e^{itA_e}\, J_e\, e^{-itA_0}, \tag{7.111}$$

provided that strong limits exist on the strong topology on $\mathcal{H}_e$. Note that since by the Rellich local compactness theorem

$$s - \lim_{t\to\pm\infty} \chi_{\Omega_i}\, e^{-itA_0} = 0, \tag{7.112}$$

it follows that

$$W_e^{\pm} = s - \lim_{t\to\pm\infty} e^{itA_T}\, J\, e^{-itA_0} = W_T^{\pm}. \tag{7.113}$$

Then the $W_e^{\pm}$ exist and are unitary from $\mathcal{H}_0$ onto $\mathcal{H}_{ac}(A_e)$.

Moreover,

$$S_e = W_e^{+*}\, W_e^{-} = W_T^{+*}\, W_T^{-} = S_T. \tag{7.114}$$

Then S_e is unitary and the results in Lemmas 7.3, 7.4, and Remarks 7.5 to 7.7 apply to $W_e^{\pm}$ and S_e.

3

Propagation of Electromagnetic Waves

§1. The Unperturbed Electromagnetic Propagator

The propagation of electromagnetic waves in three dimensional dielectric wave guides is described by the Maxwell system of equations

$$\nabla \times \overline{E} = -\mu_0(z) \frac{\partial \overline{H}}{\partial t}, \tag{1.1}$$

$$\nabla \times \overline{H} = \epsilon_0(z) \frac{\partial \overline{E}}{\partial t}, \tag{1.2}$$

$$\nabla \cdot (\epsilon_0(z)\overline{E}) = 0, \qquad \nabla \cdot (\mu_0(z)\overline{H}) = 0, \tag{1.3}$$

where $x = (x_1, x_2) \in \mathbf{R}^2$, $z \in \mathbf{R}$, $t \in \mathbf{R}$, and $\overline{E}(x,z,t) = (E_1(x,z,t), E_2(x,z,t), E_3(x,z,t))$, $\overline{H}(x,z,t) = (H_1(x,z,t), H_2(x,z,t), H_3(x,z,t))$, are the functions from $\mathbf{R}^4$ into $\mathbf{R}^3$ that correspond respectively to the electric and magnetic fields. By ∇ we denote the gradient operator

$$\nabla = \left(\frac{\partial}{\partial x_1}, \frac{\partial}{\partial x_2}, \frac{\partial}{\partial z}\right), \tag{1.4}$$

and $\nabla \times$ and $\nabla \cdot$ are respectively the curl and divergence operators. $\epsilon_0(z)$ and $\mu_0(z)$ are real valued measurable and bounded functions defined on $\mathbf{R}$ that represent respectively the electric and magnetic susceptibilities. As in Chapter 2 the derivatives in (1.1) to (1.3) are in distribution sense.

We assume that

$$0 < c_m \leq \epsilon_0(z) \leq c_M, \quad 0 < c_m \leq \mu_0(z) \leq c_M, \tag{1.5}$$

for some positive constants c_m and c_M.

Furthermore, we assume that the following limits exist

$$\epsilon_\pm = \lim_{z\to\pm\infty} \epsilon_0(z), \tag{1.6}$$

$$\mu_\pm = \lim_{z\to\pm\infty} \mu_0(z), \tag{1.7}$$

and that

$$\pm \int_0^{\pm\infty} |\epsilon_0(z) - \epsilon_\pm|\, dz < \infty, \tag{1.8}$$

$$\pm \int_0^{\pm\infty} |\mu_0(z) - \mu_\pm|\, dz < \infty. \tag{1.9}$$

This defines a general class of profiles containing most of the particular cases that appear in the applications. It was introduced in the acoustic case in Wilcox 1984.

We consider finite energy solutions to the system (1.1) to (1.3), that is to say solutions that satisfy

$$\int \left(\sum_{i=1}^{3} \left(\epsilon_0(z)\, |E_i|^2 + \mu_0(z)\, |H_i|^2\right) \right) dx dz < \infty. \tag{1.10}$$

We will formulate this problem in Hilbert space. To do so we consider $\mathbf{C}^3$ valued solutions.

Since it is customary in the physics and engineering literature (see for example Marcuse 1974) to work with T.E. (transverse electric) and T.M. (transverse magnetic) modes, we will use the same type of generalized eigenfunctions.

In order to do so we will write the Maxwell system in an equivalent form.

We first define the Hilbert space $\mathcal{H}_0$, of measurable square integrable $\mathbf{C}^6$ valued functions on $\mathbf{R}^3$ with scalar product

$$(u, v)_{\mathcal{H}_0} = \int \left[\epsilon_0 u_1 \overline{v}_1 + \mu_0 u_2 \overline{v}_2 + \mu_0 u_3 \overline{v}_3 + \mu_0 u_4 \overline{v}_4 + \epsilon_0 u_5 \overline{v}_5 + \epsilon_0 u_6 \overline{v}_6\right] dx dz. \tag{1.11}$$

We define in $\mathcal{H}_0$ the following selfadjoint operator H_0

$$H_0 u = G_0^{-1} B u, \tag{1.12}$$

$$D(H_0) = \{\, u \in \mathcal{H}_0 \ :\ Bu \in \mathcal{H}_0 \,\}, \tag{1.13}$$

where

$$G_0(z) = \begin{bmatrix} \epsilon_0 & 0 & 0 & 0 & 0 & 0 \\ 0 & \mu_0 & 0 & 0 & 0 & 0 \\ 0 & 0 & \mu_0 & 0 & 0 & 0 \\ 0 & 0 & 0 & \mu_0 & 0 & 0 \\ 0 & 0 & 0 & 0 & \epsilon_0 & 0 \\ 0 & 0 & 0 & 0 & 0 & \epsilon_0 \end{bmatrix}, \tag{1.14}$$

and

$$B = i \begin{bmatrix} 0 & -\frac{\partial}{\partial z} & \frac{\partial}{\partial x_2} & 0 & 0 & 0 \\ -\frac{\partial}{\partial z} & 0 & 0 & 0 & 0 & \frac{\partial}{\partial x_1} \\ \frac{\partial}{\partial x_2} & 0 & 0 & 0 & -\frac{\partial}{\partial x_1} & 0 \\ 0 & 0 & 0 & 0 & \frac{\partial}{\partial z} & -\frac{\partial}{\partial x_2} \\ 0 & 0 & -\frac{\partial}{\partial x_1} & \frac{\partial}{\partial z} & 0 & 0 \\ 0 & \frac{\partial}{\partial x_1} & 0 & -\frac{\partial}{\partial x_2} & 0 & 0 \end{bmatrix}. \tag{1.15}$$

The derivatives of B are in distribution sense.

By explicit computation one checks that zero is an infinite dimensional eigenvalue of H_0, and that the orthogonal complement in $\mathcal{H}_0$ to the kernel of H_0 is given by

$$(kernel(H_0))^{\perp} = \Big\{ v \in \mathcal{H}_0 \ : \ \frac{\partial}{\partial x_1}\ \epsilon_0 v_1 + \frac{\partial}{\partial x_2}\ \epsilon_0 v_5 + \frac{\partial}{\partial z}\ \epsilon_0 v_6 = 0 \ : \ \frac{\partial}{\partial x_1}\ \mu_0 v_4 + \frac{\partial}{\partial x_2}\ \mu_0 v_2 + \frac{\partial}{\partial z}\ \mu_0 v_3 = 0 \Big\}. \tag{1.16}$$

Under the identification $v_1 = E_1,\ v_2 = H_2,\ v_3 = H_3,\ v_4 = H_1,\ v_5 = E_2$, and $v_6 = E_3$ the Maxwell system (1.1) to (1.3) is equivalent to the Schrödinger type equation

$$i\ \frac{\partial}{\partial t}\ v = H_0 v, \tag{1.17}$$

for $v(t, x, z)$ in the H_0-invariant subspace $(kernel(H_0))^\perp$.

In this way the problem of studying the propagation of electromagnetic waves in three dimensional dielectric wave guides is formulated as the problem of the spectral theory of the selfadjoint operator H_0 in $\mathcal{H}_0$, specifically, the problem of generalized eigenfunctions expansion for H_0.

We will prove a theorem in eigenfunctions expansion for the operator H_0. We prove that the generalized Fourier maps define a unitary operator. This implies in particular that the sets of T.E. and T.M. modes are complete. This also gives us a spectral representation for the restriction of H_0 to the orthogonal complement in $\mathcal{H}_0$ of its kernel.

This result is obtained for the general class of profiles above by first using the fact that $\epsilon_0(z)$ and $\mu_0(z)$ are independent of the x variables to reduce the problem to that of a direct integral whose fiber is an ordinary differential operator in the z variable. This goes back to the work of Wilcox 1984 in the acoustic case.

In a second step a unitary rotation operator is introduced that reduces the 6×6 fiber ordinary differential operator to a direct sum of two 3×3 operators. This statement is analogous to the well known fact that the solutions independent of one of the transversal coordinates x_1, x_2, decompose both as a T.E. solution and a T.M. solution.

The crucial point of the proof is the realization that each of the 3×3 ordinary differential operators is, when reduced to the orthogonal complement of its kernel, unitarily equivalent to an acoustic problem (more precisely to a 2×2 system that corresponds to the spatial part of an acoustic problem reduced to first order in time).

The unitary equivalence permits us to obtain our eigenfunctions expansion theorem in a simple and elegant way as a consequence of the results in the acoustic case obtained in Wilcox 1984, and of functional calculus of selfadjoint operators.

This unitary reduction to the acoustic case in three dimensions appears to be new and is of independent interest, although the existence of an unitary reduction to the acoustic case in two dimensions is well known. See Wilcox 1976b on this point.

For Ω any open set in $\mathbf{R}^n$ we denote by $L^{2,6}(\Omega)$ the Hilbert space consisting of all measurable square integrable $\mathbf{C}^6$ valued functions defined on Ω, with scalar product

$$\left(u, v\right)_{L^{2,6}(\Omega)} = \sum_{i=1}^{6} \int_\Omega u_i(x)\overline{v}_i(x)dx, \tag{1.18}$$

for $u(x) = (u_1(x),\ u_2(x), \cdots, u_6(x))$, $\quad v(x) = (v_1(x),\ v_2(x), \cdots, u_6(x))$, in $L^{2,6}(\Omega)$.

Let $\mathcal{F}$ denotes the unitary operator of Fourier transform in $L^{2,6}(\mathbf{R}^2)$

$$(\mathcal{F}f)(k) = s - \lim_{M\to\infty} \frac{1}{(2\pi)} \int\limits_{|x|\leq M} e^{-ik\cdot x} f(x)dx, \tag{1.19}$$

where $k = (k_1, k_2) \in \mathbf{R}^2$, $x = (x_1, x_2) \in \mathbf{R}^2$, and $k \cdot x = k_1 x_1 + k_2 x_2$. The limit exists in the strong topology in $L^{2,6}(\mathbf{R}^2)$. We also denote by $\mathcal{F}$ the operator $\mathcal{F} \otimes I$ acting on $\mathcal{H}_0$ that consists in taking the partial Fourier transform on the (x_1, x_2) variables and is the identity on z. $\mathcal{F}$ is an unitary operator on $\mathcal{H}_0$. We denote

$$H_{\mathcal{F}} = \mathcal{F} \, H_0 \, \mathcal{F}^*. \tag{1.20}$$

Let $\mathcal{L}$ be the Hilbert space of measurable square integrable, $\mathbf{C}^6$ valued functions on $\mathbf{R}$ with scalar product

$$(u, v)_{\mathcal{L}} = \int \left[\epsilon_0 u_1 \overline{v}_1 + \mu_0 u_2 \overline{v}_2 + \mu_0 u_3 \overline{v}_3 + \mu_0 u_4 \overline{v}_4 + \epsilon_0 u_5 \overline{v}_5 + \epsilon_0 u_6 \overline{v}_6 \right] dz, \tag{1.21}$$

for $u(z) = (u_1(z), \cdots, u_6(z))$, $v(z) = (v_1(z), \cdots, v_6(z))$ in $\mathcal{L}$.

Clearly $\mathcal{H}_0$ is a direct integral over $\mathbf{R}^2$ with constant fiber equal to $\mathcal{L}$,

$$\mathcal{H}_0 = \oplus \int_{\mathbf{R}^2} \mathcal{L} \, dk_1 \, dk_2. \tag{1.22}$$

$H_{\mathcal{F}}$ decomposes along this direct integral as

$$H_{\mathcal{F}} = \oplus \int H(k) dk, \tag{1.23}$$

where $k = (k_1, \ k_2) \in \mathbf{R}^2$, and $H(k)$ is the selfadjoint operator in $\mathcal{L}$ defined as

$$H(k)v = G_0^{-1}(z) B(k) v, \tag{1.24}$$

$$D(H(k)) = \{ v \in \mathcal{L} \, : \, B(k)v \in \mathcal{L} \}, \tag{1.25}$$

where $G_0(z)$ is as in (1.14) and

$$B(k)v = i \begin{bmatrix} 0 & -\frac{d}{dz} & ik_2 & 0 & 0 & 0 \\ -\frac{d}{dz} & 0 & 0 & 0 & 0 & ik_1 \\ ik_2 & 0 & 0 & 0 & -ik_1 & 0 \\ 0 & 0 & 0 & 0 & \frac{d}{dz} & -ik_2 \\ 0 & 0 & -ik_1 & \frac{d}{dz} & 0 & 0 \\ 0 & ik_1 & 0 & -ik_2 & 0 & 0 \end{bmatrix} v \ . \tag{1.26}$$

The usual procedure in physics literature is to perform a rotation of coordinates that brings the x_2 axis into the direction defined by k making the solution x_1 independent. Since $\overline{E}$ and $\overline{H}$ behave as vectors under rotation in the x plane this corresponds to the unitary transformation on $\mathcal{L}$ defined by the following matrix

$$U(k) = \begin{bmatrix} \frac{k_2}{|k|} & 0 & 0 & 0 & -\frac{k_1}{|k|} & 0 \\ 0 & \frac{k_2}{|k|} & 0 & \frac{k_1}{|k|} & 0 & 0 \\ 0 & 0 & 1 & 0 & 0 & 0 \\ 0 & -\frac{k_1}{|k|} & 0 & \frac{k_2}{|k|} & 0 & 0 \\ \frac{k_1}{|k|} & 0 & 0 & 0 & \frac{k_2}{|k|} & 0 \\ 0 & 0 & 0 & 0 & 0 & 1 \end{bmatrix}, \tag{1.27}$$

$k \neq 0$. Since $U(k)$ is an unitary matrix and $G_0(z)$ commutes with $U(k)$, the function $v \to U(k)v$ defines an unitary operator on $\mathcal{L}$, that we also denote by $U(k)$. It is easily checked that for $k \neq 0$

$$U(k)H(k)U^{-1}(k) = H(0, |k|), \tag{1.28}$$

where $|k| = (k_1^2 + k_2^2)^{1/2}$.

We denote by $\mathcal{L}_E$ the Hilbert space of measurable, square integrable, $\mathbf{C}^3$ valued functions on $\mathbf{R}$ with scalar product

$$(u, v)_E = \int \left[\epsilon_0 \; u_1\overline{v}_1 + \mu_0 \; u_2\overline{v}_2 + \mu_0 \; u_3\overline{v}_3\right] dz, \tag{1.29}$$

for

$$u = (u_1, u_2, u_3), \; (v_1, v_2, v_3) \in \mathcal{L}_E. \tag{1.30}$$

We similarly define $\mathcal{L}_M$ with scalar product

$$(u, v)_M = \int \left[\mu_0 \; u_1\overline{v}_1 + \epsilon_0 \; u_2\overline{v}_2 + \epsilon_0 u_3\overline{v}_3\right] dz. \tag{1.31}$$

Clearly

$$\mathcal{L} = \mathcal{L}_E \oplus \mathcal{L}_M. \tag{1.32}$$

We define for $\rho \in (0, \infty)$ the selfadjoint operator $H_E(\rho)$ in $\mathcal{L}_E$ as follows

$$H_E(\rho)v = G_E^{-1} Dv, \tag{1.33}$$

$$D(H_E(\rho)) = \{v \in \mathcal{L}_E \; : \; D\, v \in \mathcal{L}_E\}, \tag{1.34}$$

where

$$D = i \begin{bmatrix} 0 & -\frac{d}{dz} & i\rho \\ -\frac{d}{dz} & 0 & 0 \\ i\rho & 0 & 0 \end{bmatrix} , \tag{1.35}$$

$$G_E = \begin{bmatrix} \epsilon_0 & 0 & 0, \\ 0 & \mu_0 & 0, \\ 0 & 0 & \mu_0 \end{bmatrix} . \tag{1.36}$$

We similarly define for $\rho \in (0, \infty)$, $H_M(\rho)$ as a selfadjoint operator in $\mathcal{L}_M$ as follows:

$$H_M(\rho)v = -G_M^{-1}\, Dv , \tag{1.37}$$

$$D(H_M(\rho)) = \{v \in \mathcal{L}_M \ : \ D\, v \in \mathcal{L}_M\} , \tag{1.38}$$

where

$$G_M = \begin{bmatrix} \mu_0 & 0 & 0 \\ 0 & \epsilon_0 & 0 \\ 0 & 0 & \epsilon_0 \end{bmatrix} . \tag{1.39}$$

We have that

$$H(0, |k|) = H_E(|k|) \,\oplus\, H_M(|k|) . \tag{1.40}$$

The fact that $H(0, |k|)$ decomposes as a direct sum of two operators is analogous to the standard statement that the solutions to Maxwell's equations that are independent of x_1 decompose into T.E. and T.M. modes.

We introduce now the equivalent acoustic problem.

We denote by $\mathcal{L}_{\epsilon_0}$ the Hilbert space of measurable, square integrable, complex valued functions on $\mathbf{R}$ with scalar product

$$(f, g)_{\epsilon_0} = \int f(z)\ \overline{g}(z)\epsilon_0(z)dz . \tag{1.41}$$

We similarly define $\mathcal{L}_{\mu_0}$ with scalar product

$$(f, g)_{\mu_0} = \int f(z)\ \overline{g}(z)\ \mu_0(z)dz . \tag{1.42}$$

The reduced acoustic T.E. propagator is the following selfadjoint operator in $\mathcal{L}_{\epsilon_0}$, for $\rho > 0$

$$A_E(\rho)\phi = c_0^2(z)\left[-\mu_0 \frac{d}{dz}\frac{1}{\mu_0}\frac{d}{dz} + \rho^2\right]\phi\,, \tag{1.43}$$

$$D\Big(A_E(\rho)\Big) = \left\{\phi \in \mathcal{L}_{\epsilon_0} \;:\; \frac{d}{dz}\phi \in \mathcal{L}_{\epsilon_0} \;:\; \frac{d}{dz}\frac{1}{\mu_0}\frac{1}{dz}\phi \in \mathcal{L}_{\epsilon_0}\right\}, \tag{1.44}$$

where $c_0(z) = \Big(\epsilon_0(z)\mu_0(z)\Big)^{-1/2}$. We similarly define the reduced acoustic T.M. propagator as

$$A_M(\rho)\phi = c_0^2(y)\left[-\epsilon_0 \frac{d}{dz}\frac{1}{\epsilon_0}\frac{d}{dz} + \rho^2\right]\phi\,, \tag{1.45}$$

$$D\Big(A_M(\rho)\Big) = \left\{\phi \in \mathcal{L}_{\mu_0} \;:\; \frac{d}{dz}\phi \in \mathcal{L}_{\mu_0} \;:\; \frac{d}{dz}\frac{1}{\epsilon_0}\frac{1}{dz}\phi \in \mathcal{L}_{\mu_0}\right\}. \tag{1.46}$$

The acoustic energy Hilbert space, $\mathcal{H}_E$, is defined as the set of measurable $\mathbf{C}^2$ valued functions on $\mathbf{R}$, $\phi = (\phi_1, \phi_2)$, where $\phi_1 \in D\Big(A_E^{1/2}(\rho)\Big) \equiv H_1(\mathbf{R})$, $\phi_2 \in \mathcal{L}_{\epsilon_0}$. The scalar product in $\mathcal{H}_E$ is as follows

$$(\phi, \psi)_{\mathcal{H}_E} = \Big(\sqrt{A_E(\rho)}\,\phi_1,\;\; \sqrt{A_E(\rho)}\,\psi_1\Big)_{\epsilon_0} + (\phi_2, \psi_2)_{\epsilon_0}\,. \tag{1.47}$$

We define the following selfadjoint operator in $\mathcal{H}_E$

$$T_E(\rho)\phi = i\begin{bmatrix} 0 & I \\ -A_E(\rho) & 0 \end{bmatrix}\begin{bmatrix}\phi_1 \\ \phi_2\end{bmatrix}, \tag{1.48}$$

$$D\Big(T_E(\rho)\Big) = \left\{\phi \in \mathcal{H}_E \;:\; \phi_1 \in D\Big(A_E(\rho)\Big) \;:\; \phi_2 \in D\Big(\sqrt{A_E(\rho)}\Big)\right\}. \tag{1.49}$$

Note that $T_E(\rho)$ is the operator that is obtained by reducing to first order in time the acoustic problem

$$\frac{d^2}{dt^2}u + c_0^2\left[-\mu_0 \frac{d}{dz}\frac{1}{\mu_0}\frac{d}{dz} + \rho^2\right]u = 0\,. \tag{1.50}$$

We define the Hilbert space $\mathcal{H}_M$ and the operator $T_M(\rho)$, by replacing ϵ_0, and $A_E(\rho)$, by μ_0, and $A_M(\rho)$ in (1.47), (1.48), and (1.49).

We denote by $\mathcal{H}_{\epsilon_0} \equiv \mathcal{L}_{\epsilon_0} \oplus \mathcal{L}_{\epsilon_0}$. Let W_E be the following unitary operator from $\mathcal{H}_E$ onto $\mathcal{H}_{\epsilon_0}$:

$$W_E \begin{bmatrix} \phi_1 \\ \\ \phi_2 \end{bmatrix} = \frac{1}{\sqrt{2}} \begin{bmatrix} \sqrt{A_E(\rho)} & i \\ \\ \sqrt{A_E(\rho)} & -i \end{bmatrix} \begin{bmatrix} \phi_1 \\ \\ \phi_2 \end{bmatrix} . \tag{1.51}$$

Then

$$W_E T_E(\rho) W_E^{-1} = \begin{bmatrix} \sqrt{A_E(\rho)} & 0 \\ \\ 0 & -\sqrt{A_E(\rho)} \end{bmatrix} \equiv B_E(\rho) . \tag{1.52}$$

Similary we denote $\mathcal{H}_{\mu_0} = \mathcal{L}_{\mu_0} \oplus \mathcal{L}_{\mu_0}$. The unitary operator W_M from $\mathcal{H}_M$ onto $\mathcal{H}_{\mu_0}$ is defined as in (1.51) with $\sqrt{A_M(\rho)}$ instead of $\sqrt{A_E(\rho)}$. We have that

$$W_M T_M(\rho) W_M^{-1} = \begin{bmatrix} \sqrt{A_M(\rho)} & 0 \\ \\ 0 & -\sqrt{A_M(\rho)} \end{bmatrix} \equiv B_M(\rho) . \tag{1.53}$$

The crucial point is to realize, as we will prove, that not only are Maxwell's equations for T.E. and T.M. modes equivalent to an acoustic equation, but also that the Maxwell energy and the acoustic energy coincide. We will show that actually $T_E(\rho)$ and $-T_M(\rho)$ are unitarily equivalent to the restrictions of $H_E(\rho)$ and $H_M(\rho)$ to $\mathcal{L}_E^{\perp}(\rho) \equiv \Big(\text{kernel } H_E(\rho)\Big)^{\perp}$, and $\mathcal{L}_M^{\perp}(\rho) \equiv \Big(\text{kernel } H_M(\rho)\Big)^{\perp}$.

We have that

$$\mathcal{L}_E^{\perp}(\rho) = \left\{ v \in \mathcal{L}_E \ : \ v_2 = \frac{i}{\rho\mu_0} \frac{d}{dz} (\mu_0 \ v_3) \right\} , \tag{1.54}$$

$$\mathcal{L}_M^{\perp}(\rho) = \left\{ v \in \mathcal{L}_M \ : \ v_2 = \frac{i}{\rho\epsilon_0} \frac{d}{dz} (\epsilon_0 \ v_3) \right\} . \tag{1.55}$$

Let $V_E(\rho)$ be the following operator

$$V_E(\rho) v = \begin{bmatrix} \frac{i\mu_0}{\rho} \ v_3 \\ \\ -v_1 \end{bmatrix} . \tag{1.56}$$

It is easily checked that $V_E(\rho)$ is unitary from $\mathcal{L}_E^{\perp}(\rho)$ onto $\mathcal{H}_E$, and that

$$T_E(\rho) = V_E(\rho) H_E(\rho) V_E^{-1}(\rho) . \tag{1.57}$$

We define $V_M(\rho)$ by replacing $\mu_0(z)$ by $\epsilon_0(z)$ in (1.56). $V_M(\rho)$ is unitary from $\mathcal{L}_M^{\perp}(\rho)$ onto $\mathcal{H}_M$, and

$$-T_M(\rho) = V_M(\rho) H_M(\rho) V_M^{-1}(\rho) . \tag{1.58}$$

We define

$$U_E(\rho) = W_E V_E(\rho) , \tag{1.59}$$

and

$$U_M(\rho) = W_M V_M(\rho) . \tag{1.60}$$

Then

$$B_E(\rho) = U_E(\rho) H_E(\rho) U_E^{-1}(\rho) , \tag{1.61}$$

and

$$-B_M(\rho) = U_M(\rho) H_M(\rho) U_M^{-1}(\rho) . \tag{1.62}$$

By (1.61) and (1.62) the spectral analysis of $H_E(\rho)$, and $H_M(\rho)$, on $\mathcal{L}_E^{\perp}(\rho)$, and $\mathcal{L}_M^{\perp}(\rho)$ is reduced to the analysis of $\sqrt{A_E(\rho)}$ and $\sqrt{A_M(\rho)}$.

Without loss of generality we assume that

$$c_+ = \left(\frac{1}{\epsilon_+ \, \mu_+}\right)^{1/2} \leq c_- = \left(\frac{1}{\epsilon_- \, \mu_-}\right)^{1/2} . \tag{1.63}$$

We denote

$$c_m = \inf_{z \in \mathbf{R}} \, c(z) . \tag{1.64}$$

We recall some known facts about the spectrum of $A_E(\rho)$, $\rho > 0$ (see Chapter 3 of Wilcox 1984)

$$\sigma_e\Big(A_E(\rho)\Big) = \sigma_c\Big(A_E(\rho)\Big) = [c_+^2 \, \rho^2, \infty) , \tag{1.65}$$

$$\sigma_d\Big(A_E(\rho)\Big) \subset [c_m^2 \rho^2, \; c_+^2 \rho^2) , \tag{1.66}$$

and consists of isolated eigenvalues of multiplicity one that can only accumulate at $c_+^2 \rho^2$. We denote by $N^E(\rho) - 1$ the number of eigenvalues on $[c_m^2 \rho^2, \; c_+^2 \rho^2), 0 \leq N^E(\rho) \leq \infty$. We denote the eigenvalues by $\lambda_j^E(\rho)$, $1 \leq j < N^E(\rho)$, arranged in increasing order

$$c_m^2 \rho^2 < \lambda_1^E(\rho) < \lambda_2^E(\rho) < \cdots < c_+^2 \rho^2 . \tag{1.67}$$

$c_+^2 \rho^2$ may or may not be an eigenvalue. In Theorem 3.6, Chapter 3 of Wilcox 1984, sufficient conditions are given in order that $c_+^2 \rho^2$ is not an eigenvalue. In what follows we assume by simplicity that $c_+^2 \rho^2$ is not an eigenvalue.

In case it is, an extra term is to be added to the eigenfunction expansion. Note that $A_M(\rho)$ is obtained from $A_E(\rho)$ by replacing $\mu_0(z)$ by $\epsilon_0(z)$.

In the case of $A_M(\rho)$ we denote the eigenvalues by $\lambda_j^M(\rho)$, $1 \le j < N^M(\rho)$, and we assume as well that $c_+^2\rho^2$ is not an eigenvalue of $A_M(\rho)$.

Lemma 1.1

For all $k \in R^2 \setminus \{0\}$

$$\sigma_c\Big(H(k)\Big) = \Big(-\infty,\ -c_+|k|\Big] \cup \Big[c_+|k|,\ \infty\Big)\,, \tag{1.68}$$

$$\sigma_e\Big(H(k)\Big) = \sigma_c\Big(H(k)\Big) \cup \{0\}\,, \tag{1.69}$$

zero being an eigenvalue of infinite multiplicity.

Moreover,

$$\sigma_d\Big(H(k)\Big) \subset \Big(-c_+|k|,\ -c_m|k|\Big] \cup \Big[c_m|k|,\ c_+|k|\Big)\,, \tag{1.70}$$

and consists of eigenvalues $\lambda_{\pm,j}^E(|k|) = \pm\ \sqrt{\lambda_j^E(|k|)}$, $1 \le j < N^E(|k|)$, $\lambda_{\pm,j}^M(|k|) = \pm\ \sqrt{\lambda_j^M(|k|)}$, $1 \le j < N^M(|k|)$. They are of multiplicity one unless some $\lambda_j^E(|k|)$ coincides with some $\lambda_\ell^M(|k|)$, in which case the corresponding eigenvalue of $H(k)$ is of multiplicity 2.

Let $\phi_j^E(z,|k|)$, $1 \le j < N^E(|k|)$, and $\phi_j^M(z,|k|)$, $1 \le j < N^M(|k|)$, denote the eigenvectors of $A_E(|k|)$, and $A_M(|k|)$, normalized to 1 in $\mathcal{L}_{\epsilon_0}$ and $\mathcal{L}_{\mu_0}$ corresponding to the eigenvalue $\lambda_j^E(|k|)$, and $\lambda_j^M(|k|)$. We define

$$v_{\pm,j}^E(z,|k|) = \frac{1}{\sqrt{2}} \begin{bmatrix} \pm\, i\ \phi_j^E(z,|k|) \\ \dfrac{1}{\mu_0(z)\sqrt{\lambda_j^E(|k|)}} \quad \dfrac{\partial}{\partial z}\ \phi_j^E(z,|k|) \\ -\dfrac{i\ |k|}{\mu_0(z)\sqrt{\lambda_j^E(|k|)}} \quad \phi_j^E(z,|k|) \end{bmatrix}\,, \tag{1.71}$$

$$v_{\pm,j}^M(z,|k|) = \frac{1}{\sqrt{2}} \begin{bmatrix} \mp\, i\ \phi_j^M(z,|k|) \\ \dfrac{1}{\epsilon_0(z)\sqrt{\lambda_j^M(|k|)}} \quad \dfrac{\partial}{\partial z}\ \phi_j^M(z,|k|) \\ -\dfrac{i\ |k|}{\epsilon_0(z)\sqrt{\lambda_j^M(|k|)}} \quad \phi_j^M(z,|k|) \end{bmatrix}\,. \tag{1.72}$$

Then the eigenvector of $H(k)$, normalized to 1, corresponding to $\lambda_{\pm,j}^E(k)$, is

$$\phi_{\pm,j}^E(z,|k|) = U(-k_1,\ k_2)\ [v_{\pm,j}^E(z,|k|)\ \oplus\ 0]\,, \tag{1.73}$$

and the one corresponding to $\lambda_{\pm,j}^M(|k|)$ is

$$\phi^M_{\pm,j}(z,|k|) = U(-k_1,\ k_2)\ [0 \oplus v^M_{\pm,j}(z,|k|)]\ . \tag{1.74}$$

Proof: Since the kernel of $H(k)$ is the same set of functions as the kernel of $B(k)$ the fact that 0 is an eigenvalue of infinite multiplicity follows by a simple computation. The rest of the Lemma follows from (1.40), (1.48), (1.61), (1.62) and the results on the spectral analysis of $A_E(|k|)$, $A_M(|k|)$ in Chapter 3 of Wilcox 1984, and by functional calculus of selfadjoint operators.

Q. E. D.

We proceed now to construct a spectral representation for $H(k)$. We define for $\lambda \in \mathbf{R}$, $|\lambda| > c_-|k|$,

$$v^E_\pm(z,|k|,\lambda) = \frac{1}{\sqrt{|\lambda|}} \begin{cases} i\lambda\psi^E_\pm\ (z,|k|,\lambda^2) \\ \frac{1}{\mu_0(z)}\ \frac{\partial}{\partial z}\ \psi^E_\pm\ (z,|k|,\lambda^2) \\ -\ \frac{i}{\mu_0(z)}\ |k|\ \psi^E_\pm\ (z,|k|,\lambda^2)\ , \end{cases} \tag{1.75}$$

$$v^M_\pm(z,|k|,\lambda) = \frac{1}{\sqrt{|\lambda|}} \begin{cases} -i\lambda\psi^M_\pm\ (z,|k|,\lambda^2) \\ \frac{1}{\epsilon_0(z)}\ \frac{\partial}{\partial z}\ \psi^M_\pm\ (z,|k|,\lambda^2) \\ -\ \frac{i}{\epsilon_0(z)}\ |k|\ \psi^M_\pm\ (z,|k|,\lambda^2)\ . \end{cases} \tag{1.76}$$

We denote for $c_+|k| < |\lambda| < c_-|k|$,

$$v^E_0(z,|k|,\lambda) = \frac{1}{\sqrt{|\lambda|}} \begin{cases} i\lambda\psi^E_0\ (z,|k|,\lambda^2) \\ \frac{1}{\mu_0(z)}\ \frac{\partial}{\partial z}\ \psi^E_0\ (z,|k|,\lambda^2) \\ -\ \frac{i}{\mu_0(z)}\ |k|\ \psi^E_0\ (z,|k|,\lambda^2)\ , \end{cases} \tag{1.77}$$

$$v^M_0(z,|k|,\lambda) = \frac{1}{\sqrt{|\lambda|}} \begin{cases} -i\lambda\psi^M_0\ (z,|k|,\lambda^2) \\ \frac{1}{\epsilon_0(z)}\ \frac{\partial}{\partial z}\ \psi^M_0\ (z,|k|,\lambda^2) \\ -\ \frac{i}{\epsilon_0(z)}\ |k|\ \psi^M_0\ (z,|k|,\lambda^2)\ , \end{cases} \tag{1.78}$$

where the functions $\psi^E_\pm$, ψ^E_0, $\psi^M_\pm$, ψ^M_0, are respectively the generalized eigenfunctions of $A_E(|k|)$, and $A_M(|k|)$, defined in Chapter 3 of Wilcox

1984. Moreover we define for $k \in \mathbf{R}^2 \setminus 0$

$$\phi_\pm(z,k,\lambda) = U(-k_1,\ k_2)\ [v_\pm^E(z,|k|,\lambda)\ \oplus\ v_\pm^M(z,|k|,\lambda)]\ , \quad (1.79)$$

$$\phi_0(z,k,\lambda) = U(-k_1,\ k_2)\ [v_0^E(z,|k|,\lambda)\ \oplus\ v_0^M(z,|k|,\lambda)]\ , \quad (1.80)$$

where by $v_\pm^E \oplus v_\pm^M$, we mean the vector in $\mathbf{C}^6$ whose first three components are those of $v_\pm^E$, and whose last three components are those of $v_\pm^M$. $v_0^E \oplus v_0^M$ is similarly defined.

We denote by $\mathcal{L}^\perp(k)$ the orthogonal complement of the kernel of $H(k)$. It follows from a simple computation that

$$\mathcal{L}^\perp(k) = \{v \in \mathcal{L}\ :\ ik_1\epsilon_0 v_1 + ik_2\epsilon_0 v_5 + \frac{d}{dz}\ \epsilon_0 v_6 = 0\ :$$

$$ik_1\mu_0 v_4 + ik_2\mu_0 v_2 + \frac{d}{dz}\ \mu_0 v_3 = 0\}\ . \quad (1.81)$$

For any $k \in \mathbf{R}^2 \setminus 0$ and any Borel set $\Lambda \subset \mathbf{R}$ we denote by $E_k(\Lambda)$ the spectral projector for Λ of the restriction of $H(k)$ to $\mathcal{L}^\perp(k)$. By $\chi_N(z)$ we denote the characteristic function of $[-N, N]$.

Lemma 1.2

For every $v \in \mathcal{L}$, $\quad k \in \mathbf{R}^2 \setminus 0$ the following limits

$$\hat{v}_{\pm,k}(\lambda) = s - \lim_{N\to\infty}\ \Big(v(z),\ \chi_N(z)\phi_\pm(z,k,\lambda)\Big)_{\mathcal{L}}\ , \quad (1.82)$$

$$\hat{v}_{0,k}(\lambda) = s - \lim_{N\to\infty}\ \Big(v(z),\ \chi_N(z)\phi_0(z,k,\lambda)\Big)_{\mathcal{L}}\ , \quad (1.83)$$

exist respectively as strong limits in $L^2\Big((-\infty,\ -c_-|k|) \cup (c_-|k|,\infty)\Big)$ and $L^2\Big((-c_-|k|,\ -c_+|k|) \cup (c_+|k|,\ c_-|k|)\Big)$, and moreover they are equal to zero for $v \in \text{kernel}\Big(H(k)\Big)$. Furthermore the maps $F_{\pm,k}$, $F_{0,k}$ defined for $v \in \mathcal{L}^\perp(k)$ as

$$F_{\pm,k}v = \hat{v}_{\pm,k}\ , \quad (1.84)$$

$$F_{0,k}v = \hat{v}_{0,k}\ , \quad (1.85)$$

are partially isometric from $\mathcal{L}^\perp(k)$ onto $L^2\Big((-\infty,-c_-|k|) \cup (c_-|k|,\infty)\Big)$, and onto $L^2\Big((-c_-|k|,\ -c_+|k|) \cup (c_+|k|,\ c_-|k|)\Big)$. The adjoint operators

$F^*_{\pm,k}$, and $F^*_{0,k}$ are given by

$$F^*_{\pm,k} v = s - \lim_{\substack{N\to\infty \\ \delta\downarrow 0}} \int_{c_-|k|+\delta<|\lambda|<N} \phi_\pm(z,|k|,\lambda)v(\lambda)d\lambda \quad , \tag{1.86}$$

$$F^*_{0,k} v = s - \lim_{\delta\downarrow 0} \int_{c_+|k|+\delta<|\lambda|<c_-|k|-\delta} \phi_0(z,|k|,\lambda)v(\lambda)d\lambda \quad , \tag{1.87}$$

where the limits exist in the strong topology in $\mathcal{L}$.

Moreover for $1 \leq j < N^E(|k|)$ we define the operators $F^E_{k,\pm,j}$ from $\mathcal{L}$ onto $\mathbf{C}$.

$$F^E_{k,\pm,j} v = \left(v,\ \phi^E_{\pm,j}\right)_{\mathcal{L}} \quad . \tag{1.88}$$

We similarly define for $1 < j < N^M(|k|)$

$$F^M_{k,\pm,j} v = \left(v,\ \phi^M_{\pm,j}\right)_{\mathcal{L}} \quad . \tag{1.89}$$

Then $F^E_{k,\pm,j} v = F^M_{k,\pm,j} v = 0$, for $v \in \text{kernel}\Big(H(|k|)\Big)$, and they are partially isometric from $\mathcal{L}^\perp(k)$ onto $\mathbf{C}$

We define the following orthogonal projectors

$$P_{\pm,k} = F^*_{\pm,k}\ F_{\pm,k}\ ;\ P_{0,k} = F^*_{0,k}\ F_{0,k}\ ;$$

$$P^E_{k,\pm,j} = F^{E*}_{k,\pm,j}\ F^E_{k,\pm,j}\ ;\ P^M_{k,\pm,j} = F^{M*}_{k,\pm,j}\ F^M_{k,\pm,j} \quad . \tag{1.90}$$

Then

$$P_{+,k}\ \oplus\ P_{-,k} = E_k\Big((-\infty, -c_-|k|) \cup (c_-|k|, \infty)\Big)\ , \tag{1.91}$$

$$P_{0,k} = E_k\Big((-c_-|k|,\ -c_+|k|) \cup (c_+|k|,\ c_-|k|)\Big)\ . \tag{1.92}$$

$$E_k\Big(\{\lambda^E_{\pm,j}\}\Big) = P^E_{k,\pm,j},\ E_k\Big(\{\lambda^M_{\pm,j}\}\Big) = P^M_{k,\pm,j}\ , \tag{1.93}$$

unless for some $j, \ell, \lambda^E_{\pm,j} = \lambda^M_{\pm,\ell}$, in which case

$$E_k\Big(\{\lambda^E_{\pm,j}\}\Big) = P^E_{k,\pm,j}\ \oplus\ P^M_{k,\pm,j} \quad . \tag{1.94}$$

Furthermore,

$$F_{\pm,k}\ F^*_{\pm,k}\ =\ I \text{ in } L^2\Big((-\infty,\ -c_-|k|) \cup (c_-|k|,\ \infty)\Big) \quad , \tag{1.95}$$

$$F_{0,k}\ F^*_{0,k} = I \text{ in } L^2\Big((-c_-|k|,\ -c_+|k|) \cup (c_+|k|,\ c_-|k|)\Big), \tag{1.96}$$

$$F^E_{k,\pm,j}\ F^{E*}_{k,\pm,j} = I,\ F^M_{k,\pm,j}\ F^{M*}_{k,\pm,j} = I \text{ in } \mathbf{C} \quad . \tag{1.97}$$

Let $\hat{\mathcal{H}}(|k|)$ denote the Hilbert space

$$\begin{aligned}\hat{\mathcal{H}}(|k|) =& L^2\Big((-\infty,\ -c_-|k|) \cup (c_-|k|,\ \infty)\Big)\\ &\oplus L^2\Big((-\infty,\ -c_-|k|) \cup (c_-|k|,\ \infty)\Big)\\ &\oplus L^2\Big((-c_-|k|,\ -c_+|k|) \cup (c_+|k|,\ c_-|k|)\Big)\\ &\cdot \bigoplus_{i=1}^{N^E(|k|)-1} (\mathbf{C} \oplus \mathbf{C}) \bigoplus_{i=1}^{N^M(|k|)-1} (\mathbf{C} \oplus \mathbf{C})\,. \end{aligned} \tag{1.98}$$

Then the operator defined as

$$F_k v = \Big(F_{+,k}v,\ F_{-,k}v,\ F_{0,k}v,\ F^E_{k,+,1}v,\ F^E_{k,-,1}v, \cdots, F^M_{k,+,1}v,\ F^M_{k,-,1}v \cdots,\Big)\,, \tag{1.99}$$

for $v \in \mathcal{L}^\perp(k)$ is an unitary operator from $\mathcal{L}^\perp(k)$ onto $\hat{\mathcal{H}}(|k|)$.

Finally F_k gives a spectral representation for $H(k)$, i.e. for every $v \in D\Big(H(k)\Big) \cap \mathcal{L}^\perp(k)$

$$\begin{aligned}F_k H(k) v = \Big(&\lambda F_{+,k}v,\ \lambda F_{-,k}v,\ \lambda F_{0,k}v,\\ &\lambda^E_{+,1} F^E_{k,+,1}v,\ \lambda^E_{-,1} F^E_{k,-,1}v, \cdots,\ \lambda^M_{+,1} F^M_{k,+,1}v,\ \lambda^M_{-,1} F^M_{k,-,1}v, \cdots\Big) \end{aligned} \tag{1.100}$$

Proof: The fact that the limits in (1.82), (1.83) are zero for $v \in \text{kernel}\Big(H(k)\Big)$ follows from a simple limiting argument. By (1.40), (1.48), (1.61), and (1.62)

$$\begin{aligned}&\begin{bmatrix} \sqrt{A_E} & 0 \\ 0 & -\sqrt{A_E} \end{bmatrix} \oplus (-1) \begin{bmatrix} \sqrt{A_M} & 0 \\ 0 & -\sqrt{A_M} \end{bmatrix}\\ &= \Big(U^{-1}(k)\ (U_E^{-1} \oplus U_M^{-1})\Big)^{-1} \quad H(k)\ \Big(U^{-1}(k)\ (U_E^{-1} \oplus U_M^{-1})\Big)\,, \end{aligned} \tag{1.101}$$

where $U^{-1}(k)(U_E^{-1} \oplus U_M^{-1})$ is an unitary operator from $\mathcal{H}_{\epsilon_0} \oplus \mathcal{H}_{\mu_0}$ onto $\mathcal{L}^\perp(k)$.

Then the lemma follows from Lemma 1.1, the results on the spectral representation of $A_E(|k|)$, $A_M(|k|)$ proven in Chapter 3 of Wilcox 1984, and functional calculus of selfadjoint operators.

Q. E. D.

We proceed now to construct the generalized eigenfunctions expansion for H_0. We define

$$\Omega = \{(k, \lambda) \; : \; k \in \mathbf{R}^2 \setminus 0 \; : \; \lambda \in \mathbf{R} \; : \; |\lambda| > c_-|k|\} \; , \tag{1.102}$$

$$\Omega_0 = \{(k, \lambda) \; : \; k \in \mathbf{R}^2 \setminus 0 \; : \; \lambda \in \mathbf{R} \; : \; c_+|k| < |\lambda| < c_-|k|\} \; , \tag{1.103}$$

$$\Omega_j^E = \{k \in \mathbf{R}^2 \setminus 0 \; : \; |k| \in U_j^E\} \; , \tag{1.104}$$

$$\Omega_j^M = \{k \in \mathbf{R}^2 \setminus 0 \; : \; |k| \in U_j^M\} \; , \tag{1.105}$$

where U_j^E, U_j^M are respectively the open sets of $\rho > 0$ where $A_E(\rho)$, and $A_M(\rho)$ have a jth eigenvalue (see Wilcox 1984, Chapter 3, Section 7). These are the set of k such that $H(k)$ has the eigenvalues $\lambda_{\pm,j}^E$, and $\lambda_{\pm,j}^M$, respectively.

We denote by $L^2(\Omega)$, $L^2(\Omega_0)$, $L^2(\Omega_j^E)$, and $L^2(\Omega_j^M)$, the usual spaces of complex valued Lebesgue square integrable functions.

We define the generalized eigenfunctions of $H(k)$ as follows:
for $(k, \lambda) \in \Omega$

$$\phi_\pm(x, z, k, \lambda) = \frac{1}{(2\pi)} \, e^{ik \cdot x} \; \phi_\pm(z, k, \lambda) \; , \tag{1.106}$$

and for $(k, \lambda) \in \Omega_0$

$$\phi_0(x, z, k, \lambda) = \frac{1}{(2\pi)} \, e^{ik \cdot x} \; \phi_0(z, k, \lambda) \; . \tag{1.107}$$

Furthermore, we define

$$\phi_{\pm,j}^E(x, z, k) = \frac{e^{ik \cdot x}}{(2\pi)} \; \phi_{\pm,j}^E(z, k), \quad k \in \Omega_j^E \; , \tag{1.108}$$

$$\phi_{\pm,j}^M(x, z, k) = \frac{e^{ik \cdot x}}{(2\pi)} \; \phi_{\pm,j}^M(z, k), \quad k \in \Omega_j^M \; . \tag{1.109}$$

We denote by $\chi_N(x,z)$ the characteristic function of the ball of radius N in $\mathbf{R}^3$.

Theorem 1.3

For every $v \in \mathcal{H}_0$ the following limits

$$\tilde{v}_\pm(k,\lambda) = s - \lim_{N\to\infty} \Big(v, \chi_N \phi_\pm(x,z,k,\lambda)\Big)_{\mathcal{H}_0}, \tag{1.110}$$

$$\tilde{v}_0(k,\lambda) = s - \lim_{N\to\infty} \Big(v, \chi_N \phi_0(x,z,k,\lambda)\Big)_{\mathcal{H}_0}, \tag{1.111}$$

$$\tilde{v}^E_{\pm,j}(k) = s - \lim_{N\to\infty} \Big(v, \chi_N \phi^E_{\pm,j}(x,z,k)\Big)_{\mathcal{H}_0}, \tag{1.112}$$

$$\tilde{v}^M_{\pm,j}(k) = s - \lim_{N\to\infty} \Big(v, \chi_N \phi^M_{\pm,j}(x,z,k)\Big)_{\mathcal{H}_0}, \tag{1.113}$$

exist respectively in the strong topology in $L^2(\Omega)$, $L^2(\Omega_0)$, $L^2(\Omega^E_j)$, and $L^2(\Omega^M_j)$. Furthermore they are equal to zero if $v \in \text{kernel}(H_0)$. The operators defined as

$$F_\pm v = \tilde{v}_\pm(k,\lambda), \tag{1.114}$$

$$F_0 v = \tilde{v}_0(k,\lambda), \tag{1.115}$$

$$F^E_{\pm,j} v = \tilde{v}^E_{\pm,j}(k), \tag{1.116}$$

$$F^M_{\pm,j} v = \tilde{v}^M_{\pm,j}(k), \tag{1.117}$$

for $v \in \mathcal{H}_0$ are respectively partially isometric from $\Big(\text{kernel}(H_0)\Big)^\perp$ onto $L^2(\Omega)$, $L^2(\Omega_0), L^2(\Omega^E_j)$, and $L^2(\Omega^M_j)$. The adjoint operators are given by

$$F^*_\pm v = s - \lim_{N\to\infty} \int_{\Omega_N} \phi_\pm(x,z,k,\lambda) v(k,\lambda)\, dk d\lambda, \tag{1.118}$$

$$F^*_0 v = s - \lim_{N\to\infty} \int_{\Omega_{0,N}} \phi_0(x,z,k,\lambda) v(k,\lambda)\, dk d\lambda, \tag{1.119}$$

$$F^{E*}_{\pm,j} v = s - \lim_{N\to\infty} \int_{\Omega^E_{j,N}} \phi^E_{\pm,j}(x,z,k) v(k)\, dk, \tag{1.120}$$

$$F^{M*}_{\pm,j} v = s - \lim_{N\to\infty} \int_{\Omega^M_{j,N}} \phi^M_{\pm,j}(x,z,k) v(k)\, dk, \tag{1.121}$$

where Ω_N is any compact set, $\Omega_N \subset \Omega$, and such that its characteristic function tends to 1 almost everywhere in Ω when $N \to \infty$. The sets $\Omega_{0,N}$, $\Omega^E_{j,N}$ and $\Omega^M_{j,N}$ are similarly defined.

We denote

$$\tilde{\mathcal{H}} = L^2(\Omega) \oplus L^2(\Omega) \oplus L^2(\Omega_0) \oplus \sum_{j=1}^{N^E-1} \left(L^2(\Omega^E_j) \oplus L^2(\Omega^E_j) \right)$$

$$\oplus \sum_{j=1}^{N^M-1} \left(L^2(\Omega^M_j) \oplus L^2(\Omega^M_j) \right), \quad (1.122)$$

where $N^E = \sup_{\rho>0} N^E(\rho)$, $N^M = \sup_{\rho>0} N^M(\rho)$.

We denote by $\tilde{F}$ the operator from $\mathcal{H}_0$ onto $\tilde{\mathcal{H}}$ given by

$$\tilde{F}v = \{F_+v,\ F_-v,\ F_0v,\ F^E_{+,1}v,\ F^E_{-,1}v, \cdots,\ F^M_{+,1}v,\ F^M_{-,1}v, \cdots\}. \quad (1.123)$$

Then

$$\text{kernel } (\tilde{F}) = \text{kernel } (H_0). \quad (1.124)$$

The restriction of $\tilde{F}$ to $\left(\text{kernel}(H_0)\right)^{\perp}$ is an unitary operator from $\left(\text{kernel } (H_0)\right)^{\perp}$ onto $\tilde{\mathcal{H}}$. Furthermore it gives a spectral representation for H_0 in $\left(\text{kernel}(H_0)\right)^{\perp}$, i.e. for every $v \in D(H_0) \cap \left(\text{kernel}(H_0)\right)^{\perp}$

$$\tilde{F}\ H_0 v = \{\lambda F_+ v,\ \lambda F_- v,\ \lambda F_0 v,\ \lambda^E_{+,1} F^E_{+,1} v,\ \lambda^E_{-,1} F^E_{-,1} v, \cdots,$$

$$\lambda^M_{+,1} F^M_{+,1} v,\ \lambda^M_{-,1} F^M_{-,1} v, \cdots\}. \quad (1.125)$$

Proof: The existence of the limits in (1.110) to (1.113) follow from Lemma 1.2 and from the unitarity of the Fourier transform in the x variables as an operator on $L^{2,6}(\mathbf{R}^2)$. Moreover it follows from Theorem XIII.85 of Reed and Simon 1978 and Lemma 1.2 that for any $a, b \subset (0, \infty)$ or $a, b \subset (-\infty, 0)$, and any $u, v \in \left(\text{kernel}(H_0)\right)^{\perp}$

$$\left(u, E_{H_0}(a,b)v\right) = \int_{\Omega} \chi_{(a,b)}(\lambda)\ \Big[\tilde{u}_+(k,\lambda)\overline{\tilde{v}}_+(k,\lambda) + \tilde{u}_-(k,\lambda)\overline{\tilde{v}}_-(k,\lambda) +$$

$$+ \tilde{u}_0(k,\lambda)\overline{\tilde{v}}_0(k,\lambda)\Big]\, dk d\lambda + \int \Bigg[\sum_{j=1}^{N^E-1} \Big(\chi_{(a,b)}\left(\lambda^E_{+,j}(k)\right)\ \tilde{u}^E_{+,j}(k)\ \overline{\tilde{v}}^E_{+,j}(k) +$$

$$\chi_{(a,b)}\Big(\lambda^E_{-,j}(k)\Big)\, \tilde{u}^E_{-,j}(k)\, \overline{\tilde{v}}^E_{-,j}(k)\Big) +$$

$$+ \sum_{j=1}^{N^M-1} \Big(\chi_{(a,b)}\Big(\lambda^M_{+,j}(k)\Big)\, \tilde{u}^M_{+,j}(k)\, \overline{\tilde{v}}^M_{+,j}(k) +$$

$$+ \chi_{(a,b)}\Big(\lambda^M_{-,j}(k)\Big)\, \tilde{u}^M_{-,j}(k)\, \overline{\tilde{v}}^M_{-,j}(k)\Big)\Big]\, dk\,. \tag{1.126}$$

where $\chi_{(a,b)}(\lambda)$ is the characteristic function of the interval (a, b), and $E_{H_0}(\lambda)$ is the spectral family of H_0.

At this point the theorem follows from the argument given in Wilcox 1984, Chapter 3, Section 9.

Q. E. D.

Corollary 1.4

The only eigenvalue of H_0 is zero and it has infinite multiplicity. The restriction of H_0 to $\Big(\text{kernel}(H_0)\Big)^{\perp}$ is an absolutely continuous operator whose spectrum is $\mathbf{R}$.

Proof: As we already remarked it follows by explicit computation that zero is an infinite dimensional eigenvalue of H_0. The rest of the Corollary follows from Theorem 1.3.

Q. E. D.

§2. The Limiting Absorption Principle for the Unperturbed Electromagnetic Propagator

In this section we consider the case of an electromagnetic slab wave guide, namely when for some $h > 0$,

$$\epsilon_0(z) = \begin{cases} \epsilon_+, & h \le z < \infty\,, \\ \epsilon_h, & 0 \le z < h\,, \\ \epsilon_-, & z < 0\,, \end{cases} \tag{2.1}$$

$$\mu_0(z) = \begin{cases} \mu_+, & h \le z < \infty\,, \\ \mu_h, & 0 \le z < h\,, \\ \mu_-, & z < 0\,, \end{cases} \tag{2.2}$$

for some positive constants ϵ_+, ϵ_h, ϵ_-, μ_+, μ_h, μ_-, and h. We assume, without loosing generality, that $c_+ = 1/\sqrt{\epsilon_+\mu_+} \le c_- = 1/\sqrt{\epsilon_-\mu_-}$. We also assume that $c_h = 1/\sqrt{\epsilon_h\mu_h} < c_+$. In this case the reduced propagator $H(k)$ has non zero eigenvalues, and H_0 has guided waves. As in the acoustic case it is the interesting one for the applications. In the case $c_h \ge c_+$ there are no guided waves, and the problem is simpler.

We will construct appropriate trace maps that will allow us to prove the limiting absorption principle.

In order to introduce the trace maps we first reformulate Theorem 1.3 in terms of polar coordinates.

We define the sets Ω, Ω_0, Ω_j^E, and Ω_j^M, $1, 2, 3, \cdots$, as in (1.102) to (1.105).

As in Section 1 of Chapter 2 we define $S_c = S_+ \cup S_0 \cup S_-$, where

$$S_+ = \left\{ \omega = (\omega_0, \overline{\omega}) \in S^2_{\frac{1}{c_+}} \;:\; \omega_0 > a|\overline{\omega}| \right\}, \tag{2.3}$$

$$S_0 = \left\{ \omega = (\omega_0, \overline{\omega}) \in S^2_{\frac{1}{c_+}} \;:\; 0 < \omega_0 < a|\overline{\omega}| \right\}, \tag{2.4}$$

$$S_- = \left\{ \omega = (\omega_0, \overline{\omega}) \in S^2_{\frac{1}{c_-}} \;:\; \omega_0 < 0 \right\}, \tag{2.5}$$

where $a = \left(\frac{c_-^2}{c_+^2} - 1\right)^{1/2}$, $\omega = (\omega_0, \overline{\omega})$, and $|\overline{\omega}| = (\omega_1^2 + \omega_2^2)^{1/2}$. We define the map χ_+ from Ω onto $(\mathbf{R} \setminus 0) \otimes S_+$ as

$$\chi_+(k, \lambda) = (\lambda, \omega_+(k, \lambda)) \in (\mathbf{R} \setminus 0) \otimes S_+\,, \tag{2.6}$$

where $\omega_+(k, \lambda) = (\gamma_+, \overline{\omega})$, with

$$\gamma_+(k, \lambda) = \left(\frac{1}{c_+^2} - |\overline{\omega}|^2\right)^{1/2}, \tag{2.7}$$

$$\overline{\omega}(k, \lambda) = \frac{k}{|\lambda|} \,. \tag{2.8}$$

χ_+ is an analytic transformation from Ω onto $(\mathbf{R} \setminus 0) \otimes S_+$.

We similarly define

$$\chi_0(k, \lambda) = (\lambda, \omega_+(k, \lambda)) \in (\mathbf{R} \setminus 0) \otimes S_0 \,, \tag{2.9}$$

where $\omega_+(k, \lambda)$ is as in (2.7), (2.8). χ_0 is an analytic map from Ω_0 onto $(\mathbf{R} \setminus 0) \otimes S_0$. Finally we define χ_- from Ω onto $(\mathbf{R} \setminus 0) \otimes S_-$ as

$$\chi_-(k, \lambda) = (\lambda, \omega_-) \,, \tag{2.10}$$

where $\omega_- = (-\gamma_-, \overline{\omega})$,

$$\gamma_-(k, \lambda) = \left(\frac{1}{c_-^2} - |\overline{\omega}|^2\right)^{1/2} , \tag{2.11}$$

$$\overline{\omega} = \frac{k}{|\lambda|} \,. \tag{2.12}$$

χ_- is an analytic map from Ω onto $(\mathbf{R} \setminus 0) \otimes S_-$. Note the slight difference in the definition of χ_+, χ_0, and χ_- here and in Section 1 of Chapter 2.

We denote

$$\phi_0(x, z, \lambda, \omega) = |\lambda|(c_+\gamma_+)^{1/2} \; \phi_+(x, z, k, \lambda), \; (k, \lambda) = \chi_+^{-1}(\lambda, \omega) \,, \tag{2.13}$$

for $(\lambda, \omega) \in (\mathbf{R} \setminus 0) \otimes S_+$,

$$\phi_0(x, z, \lambda, \omega) = |\lambda|(c_+\gamma_+)^{1/2} \; \phi_0(x, z, k, \lambda), \; (k, \lambda) = \chi_0^{-1}(\lambda, \omega) \,, \tag{2.14}$$

for $(\lambda, \omega) \in (\mathbf{R} \setminus 0) \otimes S_0$, and

$$\phi_0(x, z, \lambda, \omega) = |\lambda|(c_-\gamma_-)^{1/2} \; \phi_-(x, z, k, \lambda), \; (k, \lambda) = \chi_-^{-1}(\lambda, \omega), \tag{2.15}$$

for $(\lambda, \omega) \in (\mathbf{R} \setminus 0) \otimes S_-$.

In the case of the profiles (2.1) and (2.2) the equations for eigenvalues and eigenfunctions for the reduced acoustic propagators $A_E(|k|)$ and $A_M(|k|)$ have explicit solutions. Also the generalized eigenfunctions can be explicitly computed (see Appendix 1). We state the results for $A_E(|k|)$. The case of $A_M(|k|)$ is obtained by replacing $\epsilon_0(z)$ by $\mu_0(z)$. $A_E(|k|)$ is the operator

$$A_E(|k|)\phi = c_0^2(z) \left[-\mu_0 \, \frac{d}{dz} \, \frac{1}{\mu_0} \, \frac{d}{dz} + |k|^2\right] \phi \,, \tag{2.16}$$

with the domain (1.44) in $\mathcal{L}_{\epsilon_0}$.

The continuous spectrum of $A_E(|k|)$ consists of $[c_+^2\ |k|^2,\ \infty)$ and it is absolutely continuous. The point spectrum consists of a finite number of eigenvalues of multiplicity one and it is contained in $[c_h^2\ |k|^2,\ c_+^2\ |k|^2]$. There are numbers ρ_j^E, $j = 1, 2, 3, \cdots$, (see Appendix 1) such that $\rho_1^E \geq 0$, $\rho_{j+1}^E > \rho_j^E$, and $\lim_{j\to\infty} \rho_j^E = \infty$, and functions $\lambda_j^E(|k|)$, $1 \leq j < \infty$, from $(\rho_j^E,\ \infty)$ into $(0, \infty)$ such that for $\rho_j^E < |k| \leq \rho_{j+1}^E$, $A_E(|k|)$ has exactly j eigenvalues of multiplicity one, that are given by $\lambda_1^E(|k|) < \lambda_2^E(|k|) < \cdots < \lambda_j^E(|k|)$.

Similarly $A_M(|k|)$ has eigenvalues $\lambda_j^M(|k|)$ defined for $|k| \in (\rho_j^M,\ \infty)$, for $j = 1, 2, 3 \cdots$.

As in the acoustic case we parametrize the generalized eigenfunctions (1.108), (1.109) in terms of $\lambda_j^E(|k|)$, and of $\lambda_j^M(|k|)$.

For $j = 1, 2, 3, \cdots$, we denote

$$O_j^E = (-\infty,\ -c_+\rho_j^E) \cup (c_+\rho_j^E,\ \infty), \tag{2.17}$$

$$O_j^M = (-\infty,\ -c_+\rho_j^M) \cup (c_+\rho_j^M,\ \infty). \tag{2.18}$$

Let $\beta_j^E(\lambda)$ and $\beta_j^M(\lambda)$ be the following functions from O_j^E onto $(\rho_j^E,\ \infty)$, and from O_j^M onto $(\rho_j^M,\ \infty)$

$$\beta_j^E(\lambda) = \rho \iff |\lambda| = \left(\lambda_j^E(\rho)\right)^{1/2}, \tag{2.19}$$

$$\beta_j^M(\lambda) = \rho \iff |\lambda| = \left(\lambda_j^M(\rho)\right)^{1/2}, \tag{2.20}$$

for $j = 1, 2, 3 \cdots$.

We denote for $\lambda \in O_j^E$, $\nu \in S_1^1$, $\quad j = 1, 2, 3, \cdots$

$$\phi_j^E(x, z, \lambda, \nu) = \left(\beta_j^E(\lambda)\beta_j'^E(\lambda)\right)^{1/2} \begin{cases} \phi_{+,j}^E(x, z, \beta_j^E(\lambda)\nu), & \lambda > 0, \\ \phi_{-,j}^E(x, z, \beta_j^E(\lambda)\nu), & \lambda < 0, \end{cases} \tag{2.21}$$

where $\phi_{+,j}^E(x, z, k)$ are defined on (1.108), and

$$\beta_j'^E(\lambda) = \frac{d}{d\lambda}\ \beta_j^E(\lambda)\,. \tag{2.22}$$

We similarly define for $\lambda \in O_j^M$, $\nu \in S_1^1$, $\quad j = 1, 2, 3, \cdots$,

$$\phi_j^M(x, z, \lambda, \nu) = \left(\beta_j^M(\lambda)\beta_j'^M(\lambda)\right)^{1/2} \begin{cases} \phi_{+,j}^M(x, z, \beta_j^M(\lambda)\nu), & \lambda > 0, \\ \phi_{-,j}^M(x, z, \beta_j^M(\lambda)\nu), & \lambda < 0, \end{cases} \tag{2.23}$$

where $\phi_{\pm,j}^M(x, z, k)$ are defined on (1.109), and

$$\beta_j'^M(\lambda) = \frac{d}{d\lambda}\ \beta_j^M(\lambda)\,. \tag{2.24}$$

We denote

$$\hat{\mathcal{H}}_0 = L^2(\mathbf{R},\ L^2(S_c))\ , \tag{2.25}$$

where $L^2(S_c)$ is the Lebesgue space of measurable complex valued functions on S_c square integrable with respect to the measure induced by Lebesgue measure on $\mathbf{R}^3$. We denote

$$\hat{\mathcal{H}}_j^E = L^2(O_j^E,\ L^2(S_1^1))\ , \tag{2.26}$$

$$\hat{\mathcal{H}}_j^M = L^2(O_j^M,\ L^2(S_1^1))\ , \tag{2.27}$$

$1 \leq j \leq \infty$. We define

$$\hat{\mathcal{H}} = \hat{\mathcal{H}}_0 \bigoplus_{j=1}^{\infty} \left[\hat{\mathcal{H}}_j^E \oplus \hat{\mathcal{H}}_j^M\right] . \tag{2.28}$$

For any $\lambda \in \mathbf{R}$ there exists j, k such that

$$c_+\ \rho_j^E < |\lambda| \leq c_+ \rho_{j+1}^E; \quad c_+\ \rho_k^M < |\lambda| \leq c_+\ \rho_{k+1}^M\ . \tag{2.29}$$

We denote

$$\hat{\mathcal{H}}(\lambda) = L^2(S_c) \bigoplus_{i=1}^{j} L^2(S_1^1) \bigoplus_{i=1}^{k} L^2(S_1^1)\ . \tag{2.30}$$

Note that

$$\hat{\mathcal{H}} = \oplus \int_{\mathbf{R}} \hat{\mathcal{H}}(\lambda) d\lambda\ . \tag{2.31}$$

Lemma 2.1

For every $v \in \mathcal{H}_0$ the following limits

$$\hat{v}_0(\lambda, \omega) = s - \lim_{N\to\infty} \Big(v, \chi_N(x, z)\ \phi_0(x, z, \lambda, \omega)\Big)_{\mathcal{H}_0}\ , \tag{2.32}$$

$$\hat{v}_j^E(\lambda, \nu) = s - \lim_{N\to\infty} \Big(v, \chi_N(x, z)\ \phi_j^E(x, z, \lambda, \nu)\Big)_{\mathcal{H}_0}\ , \tag{2.33}$$

$$\hat{v}_j^M(\lambda, \nu) = s - \lim_{N\to\infty} \Big(v, \chi_N(x, z)\ \phi_j^M(x, z, \lambda, \nu)\Big)_{\mathcal{H}_0}\ , \tag{2.34}$$

exist respectively as strong limits in $\hat{\mathcal{H}}_0$, $\hat{\mathcal{H}}_j^E$, and $\hat{\mathcal{H}}_j^M$, $1 \leq j < \infty$. Furthermore they are equal to zero if $v \in$ kernel (H_0). The operators

$$F_0 v = \hat{v}_0(\lambda, \omega)\ , \tag{2.35}$$

$$F_j^E v = \hat{v}_j^E(\lambda, \nu)\ , \tag{2.36}$$

$$F_j^M v = \hat{v}_j^M(\lambda, \nu)\ , \tag{2.37}$$

$v \in \mathcal{H}$, $1 \leq j < \infty$, are respectively partially isometric from (kernel $H_0)^{\perp}$ onto $\hat{\mathcal{H}}_0$, $\hat{\mathcal{H}}_j^E$, and $\hat{\mathcal{H}}_j^M$. The adjoint operators are given by

$$F_0^* v = s - \lim_{\substack{a \downarrow 0 \\ b \to \infty}} \int_{(a<\lambda<b) \times S_0} \phi_0(x, z, \lambda, \omega) v(\lambda, \omega) d\lambda d\omega , \qquad (2.38)$$

$$F_j^{E*} v = s - \lim_{\substack{a \downarrow c_+ \rho_j^E \\ b \to \infty}} \int_{(a<\lambda<b) \times S_1^1} \phi_j^E(x, z, \lambda, \nu) v(\lambda, \nu) d\lambda d\nu , \qquad (2.39)$$

$$F_j^{M*} v = s - \lim_{\substack{a \downarrow c_+ \rho_j^M \\ b \to \infty}} \int_{(a<\lambda<b) \times S_1^1} \phi_j^M(x, z, \lambda, \nu) v(\lambda, \nu) d\lambda d\nu . \qquad (2.40)$$

Let F be the operator from $\mathcal{H}_0$ onto $\hat{\mathcal{H}}$ given by

$$Fv = (F_0 v,\ F_1^E v,\ F_1^M v,\ F_2^E v,\ F_2^M v, \cdots) , \qquad (2.41)$$

for $v \in \mathbf{H}_0$. Then

$$\text{kernel } (F) = \text{kernel } (H_0) . \qquad (2.42)$$

The restriction of F to (kernel $(H_0))^{\perp}$ is a unitary operator from (kernel $(H_0))^{\perp}$ onto $\hat{\mathcal{H}}$. It gives a spectral representation for H_0 in (kernel $(H_0))^{\perp}$, i.e. for every $v \in D(H_0) \cap$ (kernel $(H_0))^{\perp}$

$$F\ H_0 v = (\lambda F_0 v,\ \lambda F_1^E v,\ \lambda F_1^M v,\ \lambda F_2^E v,\ \lambda F_2^M v, \cdots) . \qquad (2.43)$$

Proof: This is a reformulation of Theorem 1.3 in terms of polar coordinates. It follows by change of variables.

Q. E. D.

Note that (2.43) is equivalent to

$$F\ H_0\ F^{-1} = \lambda , \qquad (2.44)$$

the operator of multiplication by the fiber variable in the direct integral (2.31).

We proceed now to construct bounded trace operators, i.e. we will prove that we can take a "sharp frequency" λ in (2.32) to (2.34) in a locally Hölder continuous way.

Note that $\phi_0(x, z, \lambda, \omega)$ can be written in the following form

$$\phi_0(x, z, \lambda, \omega) = \frac{1}{2\pi}\, e^{i\lambda\overline{\omega}\cdot x}\, U(-\lambda\omega_1,\ \lambda\omega_2)\ [v^E(z, \lambda, \omega) \oplus v^M(z, \lambda, \omega)]\ , \tag{2.45}$$

where

$$v_E(z, \lambda, \omega) = \begin{cases} i\lambda\ \psi^E(z, \lambda, \omega) \\ \frac{1}{\mu_0(z)}\ \frac{\partial}{\partial z}\ \psi^E(z, \lambda, \omega) \\ \frac{-i}{\mu_0(z)}\ |\lambda|\ |\overline{\omega}|\ \psi^E(z, \lambda, \omega) \end{cases} \tag{2.46}$$

$$v_M(z, \lambda, \omega) = \begin{cases} -i\lambda\ \psi^M(z, \lambda, \omega) \\ \frac{1}{\epsilon_0(z)}\ \frac{\partial}{\partial z}\ \psi^M(z, \lambda, \omega) \\ \frac{-i}{\epsilon_0(z)}\ |\lambda|\ |\overline{\omega}|\ \psi^M(z, \lambda, \omega)\ , \end{cases} \tag{2.47}$$

where for $(\lambda, \omega) \in (\mathbf{R} \setminus 0) \otimes S_+$

$$\psi^E(z, \lambda, \omega) = \sqrt{|\lambda|}\ (c_+\gamma_+)^{1/2}\ \psi_+^E(z,\ |\lambda\overline{\omega}|, \lambda^2)\ , \tag{2.48}$$

for $(\lambda, \omega) \in (\mathbf{R} \setminus 0) \otimes S_0$,

$$\psi^E(z, \lambda, \omega) = \sqrt{|\lambda|}\ (c_+\gamma_+)^{1/2}\ \psi_0^E(z,\ |\lambda\overline{\omega}|, \lambda^2)\ , \tag{2.49}$$

and for $(\lambda, \omega) \in (\mathbf{R} \setminus 0) \otimes S_-$,

$$\psi^E(z, \lambda, \omega) = \sqrt{|\lambda|}\ (c_-\gamma_-)^{1/2}\ \psi_-^E(z,\ |\lambda\overline{\omega}|, \lambda^2)\ . \tag{2.50}$$

$\psi^M(z, \lambda, \omega)$ is defined as in (2.48) to (2.50) with $\psi_\pm^M(z,\ |\lambda\overline{\omega}|,\ \lambda^2)$ and $\psi_0^M(z,\ |\lambda\overline{\omega}|,\ \lambda^2)$ instead of $\psi_\pm^E(z,\ |\lambda\overline{\omega}|,\ \lambda^2)$ and $\psi_0^E(z,\ |\lambda\overline{\omega}|,\ \lambda^2)$.

The functions $\psi_\pm^E(z,\ \rho,\ \lambda^2)$, $\psi_0^E(z,\ \rho,\ \lambda^2)$, $\psi_\pm^M(z,\ \rho,\ \lambda^2)$, and $\psi_0^M(z,\ \rho,\ \lambda^2)$ are respectively the generalized eigenfunctions of $A_E(\rho)$ and $A_M(\rho)$ defined on Chapter 3 of Wilcox 1984 (see Appendix 1).

We denote by $\mathbf{C}_0^{\infty,6}(\mathbf{R}^3)$ the space of $\mathbf{C}^6$ valued functions on $\mathbf{R}^3$ that are infinitely differentiable and have compact support. For each $s \in \mathbf{R}$ we denote by $L_s^{2,6}(\mathbf{R}^n)$ the Hilbert space of measurable, $\mathbf{C}^6$ valued functions $v(x)$, $x \in \mathbf{R}^n$ such that $(1 + |x|^2)^{s/2}\ v(x) \in L^{2,6}(\mathbf{R}^n)$ with norm

$$\left\| v \right\|_{L_s^{2,6}(\mathbf{R}^n)} = \left\| (1 + |x|^2)^{s/2} v(x) \right\|_{L^{2,6}(\mathbf{R}^n)}\ . \tag{2.51}$$

Lemma 2.2

For each $s > 1/2$, and $\lambda \in \mathbf{R} \setminus 0$ there is a trace map, $T_0(\lambda)$, bounded from $L_s^{2,6}(\mathbf{R}^3)$ into $L^2(S_c)$ and such that for every $v \in C_0^{\infty,6}(\mathbf{R}^3)$

$$\Big(T_0(\lambda)v\Big)(\omega) = \Big(v,\ \phi_0(x,z,\lambda,\omega)\Big)_{\mathcal{H}_0}. \qquad (2.52)$$

Furthermore, the function $\lambda \to T_0(\lambda)$ from $\mathbf{R} \setminus 0$ into $\mathcal{B}\Big(L_s^{2,6}(\mathbf{R}^3),\ L^2(S_c)\Big)$ is locally Hölder continuous with exponent γ where $\gamma < 1$, $\gamma < (s-1/2)$, if $\lambda \neq \pm\, c_+\, \rho_j^E$, and $\lambda \neq \pm\, c_+\, \rho_j^M$, for $j = 1,2,3,\cdots$, and $\gamma < 1/2$, $\gamma < (s-1/2)$, if $\lambda = \pm\, c_+\, \rho_j^E$, or $\lambda = \pm\, c_+\, \rho_j^M$, for some $j = 1,2,3,\cdots$.

Proof: The Lemma follows by (2.45) to (2.50), and by Lemma 1.1 in Appendix 1. Note that the results and the proof of this lemma remain valid if in formulas (1.62) to (1.64) of Appendix 1 $\psi_{\pm}(y,|k|,\lambda)$ and $\psi_0(y,|k|,\lambda)$, are replaced by $\frac{d}{dy}\ \psi_{\pm}(y,|k|,\lambda)$ and $\frac{d}{dy}\ \psi_0(y,|k|,\lambda)$.

Q. E. D.

Lemma 2.3

For each $s > 1/2$ and $\lambda \in (-\infty,\ -c_+\ \rho_j^E) \cup (c_+\ \rho_j^E, \infty)$, there is a trace map $T_j^E(\lambda)$ bounded from $L_s^{2,6}(\mathbf{R}^3)$ into $L^2(S_1^1)$ for $j = 1,2,3,\cdots$, such that for every $v \in \mathbf{C}_0^{\infty,6}(\mathbf{R}^3)$

$$\Big(T_j^E(\lambda)v\Big)(\nu) = \Big(v,\ \phi_j^E(x,z,\lambda,\nu)\Big)_{\mathcal{H}_0}. \qquad (2.53)$$

Moreover, for each $s > 1/2$ and $\lambda \in (-\infty,\ -c_+\ \rho_j^M) \cup (c_+\ \rho_j^M, \infty)$, there is a trace map $T_j^M(\lambda)$ bounded from $L_s^{2,6}(\mathbf{R}^3)$ into $L^2(S_1^1)$ for $j = 1,2,3,\cdots$, such that for every $v \in \mathbf{C}_0^{\infty,6}(\mathbf{R}^3)$

$$\Big(T_j^M(\lambda)v\Big)(\nu) = \Big(v,\ \phi_j^M(x,z,\lambda,\nu)\Big)_{\mathcal{H}_0}. \qquad (2.54)$$

Moreover, the functions $\lambda \to T_j^E(\lambda)$ and $\lambda \to T_j^M(\lambda)$ from $(-\infty,\ -c_+\ \rho_j^E) \cup (c_+\ \rho_j^E, \infty)$ into $\mathcal{B}\Big(L_s^{2,6}(\mathbf{R}^3),\ L^2(S_1^1)\Big)$, and from $(-\infty,\ -c_+\ \rho_j^M) \cup (c_+\ \rho_j^M, \infty)$ into $\mathcal{B}\Big(L_s^{2,6}(\mathbf{R}^3),\ L^2(S_1^1)\Big)$,are locally Hölder continuous with exponent $\gamma < 1$, $\gamma < (s-1/2)$.

Furthermore, if for $j = 2, 3, 4, \cdots$ we extend $T_j^E(\lambda)$ and $T_j^M(\lambda)$ to $\lambda = \pm\, c_+\, \rho_j^E$, and to $\lambda = \pm\, c_+\, \rho_j^M$ by $T_j^E(\pm\, c_+\, \rho_j^E) = 0$, and $T_j^M(\pm\, c_+\, \rho_j^M) = 0$, and if $c_+ < c_-$ we also extend $T_1^E(\lambda)$ and $T_1^M(\lambda)$ by $T_1^E(\pm\, c_+\, \rho_1^E) = 0$ and $T_1^M(\pm\, c_+\, \rho_1^M) = 0$, the functions $T_j^E(\lambda)$ and $T_j^M(\lambda)$ are Hölder continuous at $\lambda = \pm\, c_+\, \rho_j^E$ and $\lambda = \pm\, c_+\, \rho_j^M$ with exponent $\gamma \leq 1/2,\ \gamma < (s - 1/2)$.

Proof: The lemma follows from (1.108), (1.109), (2.21) to (2.24), and Lemma 1.2 in Appendix 1. Note that both the results and the proof of this lemma remain true if we replace in formula (1.66) of Appendix 1 $\psi_j(y,\ \beta_j(\lambda))$, by $\frac{d}{dy}\ \psi_j(y,\ \beta_j(\lambda))$.

Q. E. D.

For $\lambda \in \mathbf{R} \setminus 0$ we define the following operators

$$B_0(\lambda)v = T_0(\lambda)\eta_{-s}v\ , \tag{2.55}$$

$$B_j^E(\lambda)v = T_j^E(\lambda)\eta_{-s}v\ , \tag{2.56}$$

$$B_j^M(\lambda)v = T_j^M(\lambda)\eta_{-s}v\ , \tag{2.57}$$

where

$$\eta_s(x, z) = (1 + |x|^2 + z^2)^{s/2}\ , \tag{2.58}$$

and $1 \leq j < \infty$. Clearly $B_0(\lambda) \in \mathcal{B}\Big(\mathcal{H}_0,\ L^2(S_c)\Big)$, $B_j^E(\lambda),\ B_j^M(\lambda) \in \mathcal{B}\Big(\mathcal{H}_0,\ L^2(S_1^1)\Big)$, and the functions $\lambda \to B_0(\lambda)$, $\lambda \to B_j^E(\lambda)$, and $\lambda \to B_j^M(\lambda)$ are locally Hölder continuous with exponent γ as in Lemmas 2.2 and 2.3. For $\lambda \in \mathbf{R} \setminus 0$ we define the operator

$$B(\lambda)v = B_0(\lambda)v \bigoplus_{j=1}^{\infty} B_j^E(\lambda)v \bigoplus_{j=1}^{\infty} B_j^M(\lambda)v\ . \tag{2.59}$$

Clearly $B(\lambda) \in \mathcal{B}\Big(\mathcal{H}_0,\ \hat{\mathcal{H}}(\infty)\Big)$ where

$$\hat{\mathcal{H}}(\infty) = L^2(S_c) \bigoplus_{j=1}^{\infty} \Big(L^2(S_1^1) \oplus L^2(S_1^1)\Big)\ , \tag{2.60}$$

and the function $\lambda \to B(\lambda) \in \mathcal{B}\Big(\mathcal{H}_0,\ \hat{\mathcal{H}}(\infty)\Big)$ is locally Hölder continuous with exponent $\gamma < 1,\ \gamma < (s - 1/2)$ if $\lambda \neq \pm\, c_+\, \rho_j^E,\ \lambda \neq \pm\, c_+\, \rho_j^M,\ 1 \leq j <$

∞, and is locally Hölder continuous with exponent $\gamma < 1/2$, $\gamma < (s-1/2)$, if $\lambda = \pm\, c_+\, \rho_j^E$, or $\lambda = \pm\, c_+\, \rho_j^M$, $1 \leq j < \infty$. Note that actually $B(\lambda)v \in \hat{\mathcal{H}}(\lambda)$, since all the other terms in (2.59) are zero. Then also $B(\lambda) \in \mathcal{B}\big(\mathcal{H}_0,\, \hat{\mathcal{H}}(\lambda)\big)$.

It follows from Lemma 2.1 that for each Borel set $\Lambda \subset \mathbf{R} \setminus 0$

$$F\, E_{H_0}(\Lambda)\eta_{-s}v = \chi_\Lambda(\lambda) B\, v \,, \tag{2.61}$$

$$F\, H_0\, E_{H_0}(\Lambda)\eta_{-s}v = \lambda\chi_\Lambda(\lambda) B\, v \,, \tag{2.62}$$

for all $v \in \mathcal{H}_0$, where $E_{H_0}(\,\cdot\,)$ denotes the spectral projectors of H_0, and $\chi_\Lambda(\lambda)$ is the characteristic function of Λ.

We denote by P^0 the projector onto the orthogonal complement to the kernel of H_0, and $\hat{E}_{H_0}(\,\cdot\,) = E_{H_0}(\,\cdot\,)\, P^0$.

For $\lambda \in \mathbf{C} \setminus \mathbf{R}$ we denote by

$$R_0(\lambda) = (H_0 - \lambda)^{-1} \,, \tag{2.63}$$

the resolvent of H_0, and

$$\hat{R}_0(\lambda) = (H_0 - \lambda)^{-1}\, P^0 \,. \tag{2.64}$$

For any $\lambda \pm i\,\delta$, $\lambda \neq 0$, $\delta \neq 0$, let I_λ be a closed interval, $I_\lambda \subset \mathbf{R} \setminus 0$, such that λ is an interior point of I_λ. Then

$$\eta_{-s}\hat{R}_0(\lambda \pm i\,\delta)\eta_{-s} = \int_{I_\lambda} \frac{1}{\rho - \lambda \mp i\epsilon}\, B^*(\rho)B(\rho)d\rho \, +$$

$$+\, \eta_{-s}\hat{E}_{H_0}(\tilde{I}_\lambda)\hat{R}_0(\lambda \pm i\delta)\eta_{-s} \,, \tag{2.65}$$

where $\tilde{I}_\lambda$ denotes the complement of I_λ.

We denote

$$\Omega_h = \mathbf{R}^3 \setminus [(x,z) \in \mathbf{R}^3 :\; z = 0, \quad \text{or} \quad z = h] \,. \tag{2.66}$$

We denote by $H_1^6(\Omega_h)$ the Sobolev space of Lebesgue square integrable $\mathbf{C}^6$ valued functions in Ω_h whose first derivatives in distribution sense are square integrable functions. The norm in $H_1^6(\Omega_h)$ is given by

$$\left\|v\right\|_{H_1^6(\Omega_h)} = \left[\left\|v\right\|^2_{L^{2,6}(\Omega_h)} + \sum_{i=1}^{2} \left\| \frac{\partial}{\partial x_i}\, v\right\|^2_{L^{2,6}(\Omega_h)} + \left\| \frac{\partial}{\partial z}\, v\right\|^2_{L^{2,6}(\Omega_h)}\right]^{1/2} . \tag{2.67}$$

Lemma 2.4: (Coerciveness)

(kernel $(H_0))^{\perp} \cap D(H_0) \subset H_1^6(\Omega_h)$, and the imbedding is continuous, i.e. for some constant **C** and all $v \in$ (kernel $(H_0))^{\perp} \cap D(H_0)$

$$\left\| v \right\|_{H_1^6(\Omega_h)} \leq C \left[\left\| H_0 v \right\|_{\mathcal{H}_0} + \left\| v \right\|_{\mathcal{H}_0} \right] . \tag{2.68}$$

Proof: Let $v \in$ (kernel $(H_0))^{\perp} \cap D(H_0)$. Then

$$v = \frac{1}{(2\pi)} \int e^{ik\cdot x} \, \hat{v}(z,k) dk \,, \tag{2.69}$$

where $\hat{v}(z,k) \in \mathcal{L}^{\perp}(k) \cap D(H(k))$, a.e. k. Furthermore $\hat{v}(z,k) = U^{-1}(k)\left[\hat{v}^E(z,|k|) \oplus \hat{v}_M(z,|k|)\right]$, with

$$\begin{aligned} &\hat{v}_E(z,|k|) \in \mathcal{L}_E^{\perp}(|k|) \cap D(H_E(|k|)), \\ &\hat{v}_M(z,|k|) \in \mathcal{L}_M^{\perp}(|k|) \cap D(H_M(|k|)) \,. \end{aligned} \tag{2.70}$$

Then by (1.54) to (1.58)

$$\hat{v}_E(z,|k|) = \begin{bmatrix} -\phi_2(z,|k|) \\ \frac{1}{\mu_0(z)} \frac{\partial}{\partial z} \phi_1(z,|k|) \\ \frac{|k|}{i\mu_0(z)} \phi_1(z,|k|) \end{bmatrix} , \tag{2.71}$$

where $\phi_1(z,|k|) \in D(A_E(|k|))$, $\phi_2(z,|k|) \in H_1(\mathbf{R})$.

Furthermore, we have that in distribution sense in Ω_h

$$\frac{\partial}{\partial z} \hat{v}_E(z,|k|) = \begin{bmatrix} -\frac{\partial}{\partial z} \phi_2(z,|k|) \\ \frac{\partial}{\partial z} \frac{1}{\mu_0(z)} \frac{\partial}{\partial z} \phi_1(z,|k|) \\ \frac{1}{i} \frac{|k|}{\mu_0(z)} \frac{\partial}{\partial z} \phi_1(z,|k|) \end{bmatrix} . \tag{2.72}$$

Moreover,

$$H_E(|k|)\hat{v}_E(z,|k|) = \begin{bmatrix} iA_E\phi_1(z,|k|) \\ \frac{i}{\mu_0(z)} \frac{\partial}{\partial z} \phi_2(z,|k|) \\ \frac{|k|}{\mu_0(z)} \phi_2(z,|k|) \end{bmatrix} , \tag{2.73}$$

and

$$\left\|A_E \phi_1\right\|^2_{\mathcal{L}_{\epsilon_0}} \geq C \int \left[\left|\frac{\partial}{\partial z} \frac{1}{\mu_0(z)} \frac{\partial}{\partial z} \phi_1\right|^2 + |k|^4 \left|\phi_1\right|^2 + |k|^2 \left|\frac{\partial}{\partial z} \phi_1\right|^2\right] dz \,. \tag{2.74}$$

Then

$$\left\|\frac{\partial}{\partial z} \hat{v}_E\right\|_{\mathcal{L}_E} \leq C \left\|H_E(|k|)\, \hat{v}_E\right\|_{\mathcal{L}_E} \,. \tag{2.75}$$

One similarly proves that

$$\left\|\frac{\partial}{\partial z} \hat{v}_M\right\|_{\mathcal{L}_M} \leq C \left\|H_M(|k|)\, \hat{v}_M\right\|_{\mathcal{L}_M} \,. \tag{2.76}$$

Then by (1.28), (1.32), (1.40) and (2.69)

$$\left\|\frac{\partial}{\partial z} v\right\|_{L^{2,6}(\Omega_h)} \leq C \left\|H_0 v\right\|_{\mathcal{H}_0} \,. \tag{2.77}$$

Furthermore in distribution sense on $\mathbf{R}^3$

$$\frac{\partial}{\partial x_j}\, v = \frac{1}{(2\pi)} \int e^{ik\cdot x}\, ik_j\, \hat{v}(z,k) dk \,. \tag{2.78}$$

As above we prove that

$$\left\|\frac{\partial}{\partial x_j} v\right\|_{L^{2,6}(\mathbf{R}^3)} \leq C \left\|H_0 v\right\|_{\mathcal{H}_0} , \tag{2.79}$$

$j = 1, 2$.

Q. E. D.

For $s \in \mathbf{R}$ we denote by $H^6_{1,s}(\Omega_h)$ the space of all $\mathbf{C}^6$ valued functions, v in Ω_h such that $\eta_s v \in H^6_1(\Omega_h)$, with norm

$$\left\|v\right\|_{H^6_{1,s}(\Omega_h)} = \left\|\eta_s v\right\|_{H^6_1(\Omega_h)} \,. \tag{2.80}$$

We denote

$$\mathbf{C}^{\pm} = \{\lambda \in \mathbf{C} \,:\, \pm\, Im\, \lambda > 0\} \,. \tag{2.81}$$

<u>Theorem 2.5</u>:

For each $\lambda \in \mathbf{R} \setminus 0$ the following limits

$$\hat{R}_{0,\pm}(\lambda \pm i\, o) = \lim_{\delta \downarrow 0}\, \hat{R}_0(\lambda \pm i\, \delta) \,, \tag{2.82}$$

exist in the uniform operator topology in $\mathcal{B}\Big(L_s^{2,6},\ H_{1,-s}^6(\Omega_h)\Big)$, $s > 1/2$.

Furthermore the functions

$$\hat{R}_{0,\pm}(\lambda) = \begin{cases} \hat{R}_0(\lambda), & Im\ \lambda \neq 0\ , \\ \hat{R}_{0,\pm}(\lambda \pm i\ 0)\ , & Im\ \lambda = 0\ , \end{cases} \tag{2.83}$$

defined for $\lambda \in \mathbf{C}^{\pm} \cup (\mathbf{R} \setminus 0)$ are analytic for $Im\ \lambda \neq 0$, and locally Hölder continuous for $\lambda \in \mathbf{R} \setminus 0$ with exponent γ satisfying $\gamma < 1$, $\gamma < (s-1/2)$, if $\lambda \neq \pm\ c_+\ \rho_j^E$, $\lambda \neq \pm\ c_+\ \rho_j^M$, $1 \leq j < \infty$, and $\gamma < 1/2$, $\gamma < (s-1/2)$, if $\lambda = \pm\ c_+\ \rho_j^E$, or $\lambda = \pm\ c_+\ \rho_j^M$, $1 \leq j < \infty$.

Furthermore

$$\eta_{-s}\hat{R}_{0,\pm}(\lambda)\eta_{-s} = P.V. \int_{I_\lambda} \frac{1}{\rho - \lambda}\ B^*(\rho)B(\rho)d\rho \pm i\ \pi B^*(\lambda)B(\lambda) +$$

$$+\ \eta_{-s}\hat{E}_{H_0}(\tilde{I}_\lambda)\hat{R}_0(\lambda)\eta_{-s}\ . \tag{2.84}$$

Proof: If the uniform topology in $\mathcal{B}(L_s^{2,6}(\mathbf{R}^3),\ H_{1,-s}^6(\Omega_h))$ is replaced by the uniform topology in $\mathcal{B}(L_s^{2,6},\ L_{-s}^{2,6})$ the theorem follows as in the proof of Theorem 2.4 in Chapter 1. The Hölder continuity follows from the Hölder continuity of $B(\lambda)$. Moreover

$$\eta_{-s}(-i)\ \frac{\partial}{\partial x_j}\ \hat{R}_0(\lambda \pm i\ \delta)\eta_{-s} = \int_{I_\lambda} \frac{\rho}{\rho - \lambda \mp i\ \delta}\ B^*(\rho)N_jB(\rho)d\rho +$$

$$+\ \eta_{-s}(-i)\ \frac{\partial}{\partial x_j}\hat{R}_0(\lambda \pm i\ \delta)\ \hat{E}_{H_0}(\tilde{I}_\lambda)\eta_{-s}\ , \tag{2.85}$$

$j = 1, 2$, where N_j is the operator from $\hat{\mathcal{H}}(\infty)$ into $\hat{\mathcal{H}}(\infty)$ given by

$$N_j\{g_0(\omega),\ g_1^E(\nu),\ g_1^M(\nu), \cdots\} = \{\omega_j g_0(\omega), \nu_j g_1^E(\nu),\ \nu_j g_1^M(\nu), \cdots,\}\ , \tag{2.86}$$

$j = 1, 2$. Furthermore

$$\eta_{-s}\ \frac{\partial}{\partial z}\hat{R}_0(\lambda \pm i\ \delta)\eta_{-s} = \int_{I_\lambda} \frac{1}{\rho - \lambda \mp i\ \delta}\ (B'(\rho))^*B(\rho)d\rho +$$

$$+\ \eta_{-s}\ \frac{\partial}{\partial z}\hat{R}_0(\lambda \pm i\ \delta)\ \hat{E}_{H_0}(\tilde{I}_\lambda)\eta_{-s}\ , \tag{2.87}$$

where $B'(\rho)$ is defined as $B(\rho)$ but with $\frac{\partial}{\partial z}\phi_0(x, z, \lambda, \omega)$, $\frac{\partial}{\partial z}\phi_j^E(x, z, \lambda, \nu)$, $\frac{\partial}{\partial z}\phi_j^M(x, z, \lambda, \nu)$, $1 \leq j < \infty$, instead of ϕ_0, ϕ_j^E, and ϕ_j^M, where the derivatives are taken in Ω_h. $B'(\rho)$ has the same properties as $B(\rho)$.

By the coerciveness Lemma 2.4 the operators

$$\frac{\partial}{\partial x_j} \hat{R}_0(\lambda \pm i\,\delta)\, \hat{E}_{H_0}(\tilde{I}_\lambda)\,, \tag{2.88}$$

$j = 1, 2$, and

$$\frac{\partial}{\partial z} \hat{R}_0(\lambda \pm i\,\delta)\, \hat{E}_{H_0}(\tilde{I}_\lambda)\,, \tag{2.89}$$

are uniformly bounded in $\mathcal{H}_0$ for all $\delta \in \mathbf{R}$, and it follows from (2.85) to (2.87) that the theorem is also true in the uniform operator topology in $\mathcal{B}(L^{2,6}_s,\ H^6_{1,-s}(\Omega_h))$.

Q. E. D.

§3. The Limiting Absorption Principle for the Perturbed Electromagnetic Propagator

The vector Maxwell equations for the propagation of electromagnetic waves in perturbed three dimensional dielectric wave guides are

$$\nabla \times \overline{E} = -\mu(x,z)\frac{\partial \overline{H}}{\partial t}, \tag{3.1}$$

$$\nabla \times \overline{H} = \epsilon(x,z)\frac{\partial \overline{E}}{\partial t}, \tag{3.2}$$

$$\nabla \cdot (\epsilon \overline{E}) = 0, \quad \nabla \cdot (\mu \overline{H}) = 0, \tag{3.3}$$

where $x = (x_1, x_2) \in \mathbf{R}^2$, $z \in \mathbf{R}$, and as in Section 1, $\overline{E}(x,z,t) = (E_1(x,z,t),\ E_2(x,z,t),\ E_3(x,z,t))$, $\overline{H}(x,z,t) = (H_1(x,z,t),\ H_2(x,z,t),\ H_3(x,z,t))$ are functions from $\mathbf{R}^4$ into $\mathbf{R}^3$ that correspond to the electric and magnetic fields. ∇ is the gradient operator (see (1.4)) and $\nabla \times$ and $\nabla \cdot$ are the curl and divergence operators. $\epsilon(x,z)$ and $\mu(x,z)$ are real valued measurable functions defined on $\mathbf{R}^3$ that represent the electric and magnetic susceptibilities. We assume that for some positive constants c_m and c_M

$$0 < c_m \le \epsilon(x,z) \le c_M, \quad 0 < c_m \le \mu(x,z) \le c_M. \tag{3.4}$$

As in Section 1 we formulate this problem in Hilbert space, and we consider complex valued solutions to (3.1) to (3.3).

We define the Hilbert space, $\mathcal{H}$, of measurable square integrable $\mathbf{C}^6$ valued functions on $\mathbf{R}^3$ with scalar product

$$(u,v)_{\mathcal{H}} = \int \Big[\epsilon u_1 \overline{v}_1 + \mu u_2 \overline{v}_2 + \mu u_3 \overline{v}_3 +$$

$$+ \quad \mu u_4 \overline{v}_4 + \epsilon u_5 \overline{v}_5 + \epsilon u_6 \overline{v}_6 \Big] dx dz. \tag{3.5}$$

Note that $\mathcal{H}_0$, $\mathcal{H}$, and $L^{2,6}(\mathbf{R}^3)$ consists of the same set of functions, and that their norms are equivalent.

Let H be the following selfadjoint operator in $\mathcal{H}$

$$Hv = G^{-1} B\, v, \tag{3.6}$$

$$D(H) = \{v \in \mathcal{H} \quad : \quad Bv \in \mathcal{H}\}, \tag{3.7}$$

where B is as in (1.15) and

$$G(x,z) = \begin{bmatrix} \epsilon(x,z) & 0 & 0 & 0 & 0 & 0 \\ 0 & \mu(x,z) & 0 & 0 & 0 & 0 \\ 0 & 0 & \mu(x,z) & 0 & 0 & 0 \\ 0 & 0 & 0 & \mu(x,z) & 0 & 0 \\ 0 & 0 & 0 & 0 & \epsilon(x,z) & 0 \\ 0 & 0 & 0 & 0 & 0 & \epsilon(x,z) \end{bmatrix}. \tag{3.8}$$

As in section 1 one checks by explicit computation that zero is an infinite dimensional eigenvalue of H, and that the orthogonal complement in $\mathcal{H}$ of the kernel of H is given by

$$(\text{kernel}(H))^{\perp} = \{v \in \mathcal{H} : \frac{\partial}{\partial x_1}\, \epsilon v_1 + \frac{\partial}{\partial x_2}\, \epsilon v_5 + \frac{\partial}{\partial z}\, \epsilon v_6 = 0 :$$

$$\frac{\partial}{\partial x_1} \mu v_4 + \frac{\partial}{\partial x_2}\, \mu v_2 + \frac{\partial}{\partial z}\, \mu v_3 = 0\}. \tag{3.9}$$

If we identify $v_1 = E_1$, $v_2 = H_2$, $v_3 = H_3$, $v_4 = H_1$, $v_5 = E_2$, and $v_6 = E_3$, the Maxwell equations (3.1) to (3.3) are equivalent to the equation

$$i\, \frac{\partial}{\partial t}\, v = Hv, \tag{3.10}$$

in the H-invariant subspace (kernel $(H))^{\perp}$.

Let L be the following unitary operator from $\mathcal{H}$ onto $\mathcal{H}_0$

$$(L\,v)(x,z) = e^{-1}(x,z)v(x,z), \tag{3.11}$$

for $v \in \mathcal{H}$, where

$$e(x,z) = G_0^{1/2}(z)\,G^{-1/2}(x,z). \tag{3.12}$$

We denote

$$\tilde{H} = L\,H\,L^{-1} = e(x,z)H_0\;e(x,z). \tag{3.13}$$

$\tilde{H}$ is selfadjoint in $\mathcal{H}_0$ with domain

$$D(\tilde{H}) = \{v \in \mathcal{H}_0\;:\;e(x,z)v \in D(H_0)\}. \tag{3.14}$$

For $\lambda \in \mathbf{C} \setminus \mathbf{R}$ we denote by

$$R(\lambda) = (H - \lambda)^{-1}, \tag{3.15}$$

and

$$\tilde{R}(\lambda) = (\tilde{H} - \lambda)^{-1}, \tag{3.16}$$

respectively the resolvents of H and $\tilde{H}$.

We also define for $\lambda \in \mathbf{C} \setminus \mathbf{R}$

$$Q_0(\lambda) = e^{-1}(\tilde{H} - \lambda)\;e^{-1}R_0(\lambda), \tag{3.17}$$

$$Q(\lambda) = (H_0 - \lambda)\;e\;\tilde{R}(\lambda)e. \tag{3.18}$$

Note that $Q_0(\lambda)$ and $Q(\lambda)$ are bounded with bounded inverse in $\mathcal{H}$, and that

$$Q(\lambda) = Q_0^{-1}(\lambda). \tag{3.19}$$

Moreover

$$\tilde{R}(\lambda) = e^{-1}\;R_0(\lambda)\;Q_0^{-1}(\lambda)\;e^{-1}. \tag{3.20}$$

We have furthermore that

$$Q_0(\lambda) = I + \lambda\;(I - e^{-2})R_0(\lambda). \tag{3.21}$$

We denote for $\lambda \in \mathbf{C}^{\pm} \cup (\mathbf{R} \setminus 0)$

$$R_{0,\pm}(\lambda) = \hat{R}_{0,\pm}(\lambda) - \frac{1}{\lambda}\;P_0, \tag{3.22}$$

where P_0 is the projector onto the kernel of H_0. Note that $P_0 \in \mathcal{B}(L_s^{2,6}(\mathbf{R}^3), L_{-s}^{2,6}(\mathbf{R}^3))$ for all $s \geq 0$.

For $\lambda \in \mathbf{C}^{\pm} \cup (\mathbf{R}^{+} \setminus 0)$ we define

$$Q_{0,\pm}(\lambda) = I + \lambda\,(I - e^{-2}) R_{0,\pm}(\lambda). \tag{3.23}$$

Let $\epsilon_0(z)$ and $\mu_0(z)$ be the profiles (2.1), (2.2) corresponding to a slab wave guide.

We assume that for some positive constants $C,\ \delta$

$$\left|\epsilon(x,z) - \epsilon_0(z)\right| \leq C\,(1 + |x| + |z|)^{-1-\delta}, \tag{3.24}$$

$$\left|\mu(x,z) - \mu_0(z)\right| \leq C\,(1 + |x| + |z|)^{-1-\delta}. \tag{3.25}$$

Then $Q_{0,\pm}(\lambda) \in \mathcal{B}(L^{2,6}_{s}(\mathbf{R}^3),\quad L^{2,6}_{-s}(\mathbf{R}^3))$, for $s = \frac{1+\delta}{2}$.

Since $R(\lambda) \in \mathcal{B}(\mathcal{H})$, also $R(\lambda) \in \mathcal{B}(L^{2,6}_{s}(\mathbf{R}^3),\quad L^{2,6}_{-s}(\mathbf{R}^3))$, $s \geq 0$, and by (3.13) and (3.20)

$$R(\lambda) = R_{0,\pm}(\lambda) Q^{-1}_{0,\pm}(\lambda)\, e^{-2}, \tag{3.26}$$

for $\lambda \in \mathbf{C}^{\pm}$. Moreover if for some $\lambda \in \mathbf{R} \setminus 0,\quad Q_{0,\pm}(\lambda)$ is invertible on $\mathcal{B}(L^{2,6}_{s}(\mathbf{R}^3))$, (3.26) defines the extension from above or below the real axis that is required by the limiting absorption principle. We will prove that $Q_{0,\pm}(\lambda)$ is invertible if and only if λ is not an eigenvalue of H.

Remark 3.1

Note that since zero is an infinite dimensional eigenvalue of H_0 the operator $(I - e^{-2}) R_{0,\pm}(\lambda)$ is not, in general, compact in $L^{2,6}_{s}(\mathbf{R}^3),\quad s = \frac{1+\delta}{2}$.

However by the second resolvent equation

$$Q_{0,\pm}(\lambda) = Q_0(\lambda_0)\,(I + K_{\pm}(\lambda)), \tag{3.27}$$

for $\lambda \in \mathbf{C}^{\pm} \cup (\mathbf{R} \setminus 0)$ and some fixed $\lambda_0 \in \mathbf{C} \setminus \mathbf{R}$, where

$$K_{\pm}(\lambda) = (\lambda - \lambda_0) Q(\lambda_0)\,(I - e^{-2}) \hat{R}_{0,\pm}(\lambda) H_0 R_0(\lambda_0). \tag{3.28}$$

By Theorem 2.5 and the Rellich local compactness theorem (see Adams 1975) $K_{\pm}(\lambda)$ are compact operators on $L^{2}_{s}(\mathbf{R}^3),\quad s = \frac{1+\delta}{2}$. Moreover since $Q_0(\lambda_0)$ is invertible it follows from (3.27) that $Q_{0,\pm}(\lambda)$ is invertible if and only if $(I + K_{\pm}(\lambda))$ is invertible.

Before we study the invertibility of $Q_{0,\pm}(\lambda)$ we state some preliminary results.

Remark 3.2

For each $s > 1/2$, $k \in \mathbf{R}^2 \setminus 0$ and $\lambda \in (-\infty, -c_-|k|) \cup (c_-|k|, \infty)$, there are trace maps $T_{\pm,k}(\lambda)$ bounded from $L_s^{2,6}(\mathbf{R})$ into $\mathbf{C}$ such that for every $v \in C_0^{\infty,6}(\mathbf{R}^3)$

$$T_{\pm,k}(\lambda)v = \Big(v,\ \phi_\pm(z,k,\lambda)\Big)_{\mathcal{L}}. \tag{3.29}$$

The function $\lambda \to T_{\pm,k}(\lambda)$ from $(-\infty, -c_-|k|) \cup (c_-|k|, \infty)$ into $\mathcal{B}(L_s^{2,6}(\mathbf{R}^3), \mathbf{C})$ are locally Hölder continuous in λ with exponent $\gamma < 1$, $\gamma < (s - 1/2)$.

Proof: It follows from the explicit evaluation of the generalized eigenfunctions $\phi_\pm(z,k,\lambda)$ (see (1.79) in Section 1 and formulas (1.50) to (1.52) in Appendix 1) that for any compact set $K \subset (-\infty, -c_-|k|) \cup (c_-|k|, \infty)$ there is a constant C such that

$$\Big|\phi_\pm(z,k,\lambda)\Big| \le C, \tag{3.30}$$

$$\Big|\phi_\pm(z,k,\lambda) - \phi_\pm(z,k,\lambda')\Big| \le C\ (1+z^2)^{1/2}\Big|\lambda - \lambda'\Big|, \tag{3.31}$$

for all $\lambda,\ \lambda' \in K$. Then for all $0 \le \gamma \le 1$,

$$\Big|\phi_\pm(z,k,\lambda) - \phi_\pm(z,k,\lambda')\Big| \le C\ (1+z^2)^{\gamma/2}\ \Big|\lambda - \lambda'\Big|^{\gamma}. \tag{3.32}$$

For $v \in \mathbf{C}_0^{\infty,6}(\mathbf{R}^3)$ we define $T_{\pm,k}(\lambda)$ by (3.29).
We have that

$$\Big|T_{\pm,k}(\lambda)v\Big| \le \Big|\Big((1+|z|^2)^{s/2}v,\ (1+z^2)^{-s/2}\ \phi_\pm(z,k,\lambda)\Big)_{\mathcal{L}}\Big| \le$$

$$\le C\ \Big\|v\Big\|_{L_s^{2,6}(\mathbf{R})}. \tag{3.33}$$

Then $T_{\pm,k}(\lambda)$ extends to a bounded operator from $L_s^{2,6}(\mathbf{R})$ into $\mathbf{C}$.
We similarly prove that $T_{\pm,k}(\lambda)$ are locally Hölder continuous using (3.32).

Remark 3.3

We define the following bounded operator from $\mathcal{L}$ into $\mathbf{C} \oplus \mathbf{C}$, for $s > 1/2$

$$B_k(\lambda) = T_{+,k}(\lambda)\ (1+z^2)^{-s/2}\ \oplus\ T_{-,k}(\lambda)\ (1+z^2)^{-s/2}. \tag{3.34}$$

For $\lambda \in \rho(H^{\perp}(k))$ let us denote by

$$R_k(\lambda) = (H^{\perp}(k) - \lambda)^{-1}, \tag{3.35}$$

the resolvent of the restriction of $H(k) + \cdot\mathcal{L}^{\perp}(k)$, that we denote $H^{\perp}(k)$. It follows from Lemma 1.2 that for $|\lambda| > c_-|k|$, $\sigma > 0$

$$(1+z^2)^{-s/2} R_k(\lambda \pm i\,\sigma)(1+z^2)^{-s/2} = \int_{I_\lambda} \frac{1}{\rho - \lambda \mp i\,\sigma} B_k^*(\rho) B_k(\rho) d\rho +$$

$$+ (1+z^2)^{-s/2} E_k(I_\lambda^{\sim}) R_k(\lambda \pm i\,\sigma)(1+z^2)^{-s/2}, \tag{3.36}$$

where I_λ is a compact inverval contained in $(-\infty,\ -c_-|k|) \cup (c_-|k|,\ \infty)$, such that λ is an interior point of I_λ.

It follows as in the proof of Theorem 2.4 in Section 2 of Chapter 2 that the following limits

$$R_k(\lambda \pm i\,0) = \lim_{\sigma \downarrow 0} R_k(\lambda \pm i\,\sigma), \tag{3.37}$$

exist in the uniform operator topology in $\mathcal{B}(L_s^{2,6}(\mathbf{R}),\ L_{-s}^{2,6}(\mathbf{R}))$, $\quad s > 1/2$, uniformly for λ in compact sets of $(-\infty,\ -c_-|k|) \cup (c_-|k|,\ \infty)$.

Moreover, the function $\lambda \to R_k(\lambda \pm i\,0)$ is locally Hölder continuous with exponent $\gamma < 1$, $\gamma < (s - 1/2)$, and

$$(1+z^2)^{-s/2} R_k(\lambda \pm i\,0)(1+z^2)^{-s/2} = P.V. \int_{I_\lambda} \frac{1}{\rho - \lambda}\, B_k^*(\rho) B_k(\rho) d\rho$$

$$\pm\, i\,\pi\, B_k^*(\lambda) B_k(\lambda) + (1+z^2)^{-s/2} E_k(\tilde{I}_\lambda) R_k(\lambda)(1+z^2)^{-s/2}. \tag{3.38}$$

Let us define the operators

$$J_k v = \int_{I_\lambda} B_k^*(\rho) v(\rho) d\rho. \tag{3.39}$$

Clearly $J_k \in \mathcal{B}(L^1(I_\lambda,\ \mathbf{C} \oplus \mathbf{C}),\ \mathcal{H}_0)$. Moreover by Lemma 1.2 for every $v \in L^2(I_\lambda,\ \mathbf{C} \oplus \mathbf{C})$

$$J_k v = (1+z^2)^{-s/2} E_k(I_\lambda) F_k^* v, \tag{3.40}$$

where we extended v to $\tilde{I}_\lambda$ by zero. It follows from (3.40) that $J_k \in \mathcal{B}(L^2(I_\lambda,\ \mathbf{C} \oplus \mathbf{C}),\ L_s^{2,6}(\mathbf{R}^3))$. Then by interpolation (see Lions and Magenes 1972, and Reed and Simon 1975, Appendix to Section IX.4), $J_k \in \mathcal{B}(L^p(I_\lambda,\ \mathbf{C} \oplus \mathbf{C}),\ L_{\epsilon_p s}^{2,6}(\mathbf{R}^3))$, where $\epsilon_p = 2\,(1 - \frac{1}{p})$, $1 \le p \le 2$.

Lemma 3.4

Let $f \in C_0^\infty(\mathbf{R})$. Then for any s, $-1 < s < 1$, $f(H_0) \in \mathcal{B}(L_s^{2,6}(\mathbf{R}^3), L_s^{2,6}(\mathbf{R}^3))$.

Proof: By duality it is enough to prove the Lemma for $0 \leq s < 1$. By a simple argument we prove that as a quadratic form on $C_0^{\infty,6}(\mathbf{R}^3) \otimes C_0^{\infty,6}(\mathbf{R}^3)$,

$$D(t) \equiv e^{itH_0}\ \eta_s - \eta_s\ e^{itH_0} = i \int_0^t e^{i\rho H_0}\ [H_0, \eta_s]\ e^{-i\rho H_0}\ e^{itH_0}\ d\rho. \quad (3.41)$$

Since the right hand side of (3.41) is a bounded operator on $\mathcal{H}_0$ it follows that $D(t)$, first defined as a quadratic form, extends to a bounded operator on $\mathcal{H}_0$. Moreover,

$$\left\| D(t) \right\|_{\mathcal{B}(\mathcal{H}_0)} \leq C\ |t|. \quad (3.42)$$

Since

$$f(H_0) = \frac{1}{\sqrt{2\pi}} \int \hat{f}(t)\ e^{itH_0} dt, \quad (3.43)$$

where $\hat{f}(t)$ denotes the Fourier transform of $f(x)$, for every $v \in C_0^{\infty,6}(\mathbf{R}^3)$

$$\eta_{-s} f(H_0) \eta_s v = \frac{\eta_{-s}}{\sqrt{2\pi}} \int_0^t \hat{f}(t) D(t) v dt + f(H_0) v. \quad (3.44)$$

Then

$$\left\| f(H_0) v \right\|_{L_s^{2,6}(\mathbf{R}^3)} \leq C \left\| v \right\|_{L_s^{2,6}(\mathbf{R}^3)}, \quad (3.45)$$

and by continuity $f(H_0)$ is a bounded operator on $L_s^{2,6}(\mathbf{R}^3)$.

Q. E. D.

Lemma 3.5

Suppose that (3.24) and (3.25) hold. Then the operator $Q_{0,\pm}(\lambda)$ is invertible on $L_s^{2,6}(\mathbf{R}^3)$, $s = \frac{1+\delta}{2}$, for $\lambda \in \mathbf{R} \setminus 0$ if and only if λ is not an eigenvalue of H.

Proof: Suppose that λ is an eigenvalue of $\tilde{H}$ with eigenvector v (note that since H and $\tilde{H}$ are unitarily equivalent they have the same eigenvalues). Then

$$0 = R_0(\lambda \pm i\,\sigma)\, e^{-1}(\tilde{H} - \lambda)v = ev + \lambda\, R_0(\lambda \pm i\,\sigma)(I - e^{-2})ev$$

$$\pm\; i\,\sigma\, R_0(\lambda \pm i\,\sigma)ev. \tag{3.46}$$

But by the spectral theorem since H_0 has no eigenvalues different from zero

$$\lim_{\sigma \downarrow 0} \sigma\, R_0(\lambda \pm i\,\sigma)ev = 0, \tag{3.47}$$

in the strong topology in $\mathcal{H}_0$. Then by (3.46) and Theorem 2.5

$$ev + \lambda\, R_{0,\pm}(\lambda)(I - e^{-2})ev = 0, \tag{3.48}$$

and denoting

$$u = (I - e^{-2})ev, \tag{3.49}$$

$$Q_{0,\pm}(\lambda)u = 0. \tag{3.50}$$

It follows that $Q_{0,\pm}(\lambda)$ is not invertible.

Suppose that $Q_{0,\pm}(\lambda)$ is not invertible. Then by Remark 3.1 and since $K_\pm(\lambda)$ are compact there is a $u \in L^{2,6}_s(\mathbf{R}^3)$, $u \neq 0$, such that

$$Q_{0,\pm}(\lambda)u = 0. \tag{3.51}$$

Note that a priori we should denote $u\pm$, since the eigenfunction could be different for $Q_{0,\pm}(\lambda)$, however, we use u by simplicity. By (3.23)

$$u = -\lambda(I - e^{-2})R_{0,\pm}(\lambda)u. \tag{3.52}$$

denote

$$\hat{u} = R_{0,\pm}(\lambda)u. \tag{3.53}$$

Note that $\hat{u} \neq 0$, because otherwise by (3.52) $u = 0$. We have that

$$\hat{u} = -\lambda\, R_{0,\pm}(\lambda)(I - e^{-2})\hat{u}. \tag{3.54}$$

Let I_λ be any compact interval, $I_\lambda \subset \mathbf{R} \setminus 0$, such that λ is an interior point of I_λ. It follows from (2.65) that

$$v = -\lambda \Big[P.V. \int_{I_\lambda} \frac{1}{\rho - \lambda}\, B^*(\rho)B(\rho)d\rho\ \pm i\pi\, B^*(\lambda)B(\lambda) +$$

$$+\ \eta_{-s}E_{H_0}(I_\lambda^\sim)R_0(\lambda)\eta_{-s}\Big](I - e^{-2})\eta_{2s}v, \tag{3.55}$$

where $v = \eta_{-s}\hat{u}$.

We have that

$$\left(v, (I - e^{-2})\eta_{2s} v\right)_{\mathcal{H}_0} = -\lambda \left[P.V. \int_{I_\lambda} \frac{1}{\rho - \lambda} \left\| B(\rho)\eta_{2s}(I - e^{-2})v \right\|^2 d\rho \right.$$

$$\pm C \; \pi \left\| B(\lambda)(I - e^{-2})\eta_{2s} v \right\|^2 +$$

$$\left. + \left(E_{H_0}(\tilde{I}_\lambda) \; R_0(\lambda)\eta_s(I - e^{-2})v, \eta_s(I - e^{-2})v\right)_{\mathcal{H}_0} \right]. \tag{3.56}$$

Then

$$B(\lambda)(I - e^{-2})\eta_{2s} v = 0, \tag{3.57}$$

and since $B(\lambda)$ is locally Hölder continuous in λ

$$v = v_\lambda + v_{\tilde{\lambda}}, \tag{3.58}$$

where

$$v_\lambda = -\lambda \int_{I_\lambda} \frac{1}{\rho - \lambda} \; B^*(\rho)B(\rho)(I - e^{-2})\eta_{2s} v, \tag{3.59}$$

$$v_{\tilde{\lambda}} = -\lambda \; \eta_{-s} E_{H_0}(\tilde{I}_\lambda) \; R_0(\lambda)(I - e^{-2})\eta_s v. \tag{3.60}$$

Moreover

$$\left\| v_{\tilde{\lambda}} \right\|_{L_s^{2,6}(\mathbf{R}^3)} \leq C \left\| v \right\|_{\mathcal{H}_0}. \tag{3.61}$$

Remark that there is a $j > 0$ such that $O_i^E \cap I_\lambda$ is not empty if $i \leq j$, and $O_i^E \cap I_\lambda$ is empty if $i \geq j + 1$. Similarly there is a k such that $O_j^M \cap I_\lambda$ is nonempty for $i \leq k$, and $O_i^M \cap I_\lambda$ is empty for $i \geq k + 1$. We denote

$$\hat{\mathcal{H}}_\lambda = \oplus \int_{I_\lambda} \hat{\mathcal{H}}(\lambda) d\lambda. \tag{3.62}$$

Note that we can identify

$$\hat{\mathcal{H}}_\lambda = L^2(I_\lambda, \; L^2(S_c)) \bigoplus_{i=1}^{j} L^2(I_{\lambda,i}^E, \; L^2(S_1^1)) \bigoplus_{i=1}^{k} L^2(I_{\lambda,i}^M, \; L^2(S_1^1)), \tag{3.63}$$

where $I_{\lambda,i}^E = I_\lambda \cap O_i^E, \quad I_{\lambda,i}^M = I_\lambda \cap O_i^M$.

Let J_λ be the following operator

$$J_\lambda(\varphi) = \int_{I_\lambda} B^*(\rho)\varphi(\rho) d\rho, \tag{3.64}$$

for $\varphi \in \hat{\mathcal{H}}_\lambda$. By (2.61)

$$J_\lambda(\varphi) = \eta_{-s} E_{H_0}(I_\lambda) F^* \varphi, \tag{3.65}$$

where we extended φ to $I_\lambda^\sim$ by zero. Then since F is unitary

$$\left\| J_\lambda \varphi \right\|_{L_s^{2,6}(\mathbf{R}^3)} \leq C \left\| \varphi \right\|_{\hat{\mathcal{H}}_\lambda}, \tag{3.66}$$

and it follows that $J_\lambda \in \mathcal{B}(\hat{\mathcal{H}}_\lambda,\ L_s^{2,6}(\mathbf{R}^3))$.

Moreover, since $B^*(\rho)$ is locally Hölder continuous, J_λ is a bounded operator from

$$L^1(I_\lambda,\ L^2(S_c)) \bigoplus_{i=1}^{j} L^1(I_{\lambda,i}^E,\ L^2(S_1^1)) \bigoplus_{i=1}^{k} L^1(I_{\lambda,i}^M,\ L^2(S_1^1)), \tag{3.67}$$

into $\mathcal{H}_0$. It follows by interpolation that J_λ is bounded from

$$L^p(I_\lambda,\ L^2(S_c)) \bigoplus_{i=1}^{j} L^p(I_{\lambda,i}^E,\ L^2(S_1^1)) \bigoplus_{i=1}^{k} L^p(I_{\lambda,i}^M,\ L^2(S_1^1)), \tag{3.68}$$

into $L_{\epsilon_p s}^{2,6}(\mathbf{R}^3)$, where $\epsilon_p = 2(1 - \frac{1}{p})$, $\quad 1 \leq p \leq 2$.

Suppose that we know that $v \in L_{s_1}^{2,6}(\mathbf{R}^3)$ for some $s_1 \geq 0$. Then by (3.57), (3.59), and (3.64)

$$v_\lambda = -\lambda\ J_\lambda \Big(\frac{1}{\rho - \lambda}\ (B(\rho) - B(\lambda))\ (I - e^{-2})\ \eta_{2s}\ \eta_{-s_1}(\eta_{s_1} v) \Big), \tag{3.69}$$

and by the Hölder continuity of $B(\rho)$

$$(B(\rho) - B(\lambda))\ (I - e^{-2})\ \eta_{2s}\ \eta_{-s_1}(\eta_{s_1} v) \in L^p(I_\lambda,\ L^2(S_c))$$

$$\bigoplus_{i=1}^{j} L^p(I_{\lambda,i}^E,\ L^2(S_1^1)) \bigoplus_{i=1}^{j} L^p(I_{\lambda,i}^M,\ L^2(S_1^1)), \tag{3.70}$$

for any $p < (1-\beta)^{-1}$, and $\beta < \min\,(1,\ s + s_1 - \frac{1}{2})$, if $\lambda \neq \pm\ c_+\ \rho_i^E$, $\quad \lambda \neq \pm\ c_+\ \rho_i^M, i = 1, 2, 3, \cdots$, and $\beta < \min\,(\frac{1}{2},\ s + s_1 - \frac{1}{2})$, if $\lambda = \pm\ c_+\ \rho_i^E$, or $\lambda = \pm\ c_+\ \rho_i^M$, for some $i = 1, 2, 3, \cdots$. Then since $v_{\tilde{\lambda}} \in L_s^{2,6}(\mathbf{R}^3)$, we know that $v \in L_{\epsilon_p s}^{2,6}(\mathbf{R}^3)$.

Suppose that $\lambda \neq \pm\ c_+\ \rho_i^E$, $\lambda \neq \pm\ c_+\ \rho_i^M$, for $i = 1, 2, 3, \cdots$. Then starting this argument with $s_1 = 0$ and iterating it a finite number of times we prove that $v \in L_s^{2,6}(\mathbf{R}^3)$, and it follows that $\hat{u} = \eta_s\ v \in \mathcal{H}_0$, and by (3.54)

$$H\hat{u} = \lambda \hat{u}, \tag{3.71}$$

and it follows that λ is an eigenvalue of H.

If $\lambda = \pm\, c_+\, \rho_i^E$, or $\lambda = \pm\, c_+\, \rho_i^M$, $i = 1, 2, 3, \cdots$, since $\beta < 1/2$, by the argument above we only prove that $\hat{u} \in L^{2,6}_{-s'}(\mathbf{R}^3)$, $0 < s' < 1/2$, and we can not conclude from (3.71) that λ is an eigenvalue of H.

However let $f \in C_0^\infty(\mathbf{R} \setminus 0)$ satisfy $f \equiv 1$ on a neighborhood of λ, and denote

$$u_1 = (I - f(H_0))\hat{u}, \tag{3.72}$$

$$u_2 = f(H_0)\hat{u}. \tag{3.73}$$

By (3.71)

$$(H_0 - \lambda)u_1 = (I - f(H_0))q_1\hat{u}, \tag{3.74}$$

and

$$(H_0 - \lambda)u_2 = f(H_0)q_1\hat{u}, \tag{3.75}$$

where

$$q_1(x, z) = \lambda\, (G\, G_0^{-1} - I). \tag{3.76}$$

We have that

$$u_1 = (H_0 - \lambda)^{-1}\, (I - f(H_0))q_1\hat{u} \in \mathcal{H}_0. \tag{3.77}$$

We will prove that also $u_2 \in \mathcal{H}_0$.

Let $\chi(\rho)$ be a function in $C_0^\infty(\mathbf{R})$ such that $\chi(\rho) \equiv 1$ in a neighborhood of $\rho = 0$, and with its support contained in the set $|\rho| < (\lambda/c_-)^2$. We define

$$\chi(-\Delta) = \mathcal{F}^*\chi(k^2)\mathcal{F}. \tag{3.78}$$

We prove as in Lemma 3.4 that $\chi(-\Delta) \in \mathcal{B}(L_s^{2,6}(\mathbf{R}^3),\ L_s^{2,6}(\mathbf{R}^3))$, for $-1 < s < 1$.

We define

$$v_1 = \chi(-\Delta)u_2, \tag{3.79}$$

$$v_2 = (I - \chi(-\Delta))u_2. \tag{3.80}$$

Since $v_1 \in L^{2,6}_{-s'}(\mathbf{R}^3)$, $0 < s' < 1/2$, for any $s > 1/2$

$$\begin{aligned} s - \lim_{\sigma \downarrow 0}\ & \eta_{-s}(H_0 - \lambda \mp i\,\sigma)^{-1}v_1 \\ &= s - \lim_{\sigma\downarrow 0}\ \eta_{-s}\ \eta_{s'}\ \sigma(H_0 - \lambda \mp\ i\,\sigma)^{-1}\ \eta_{-s'}\ v_1\ + \\ &+ \eta_{-s}(H_0 - \lambda \mp\ i\,\sigma)^{-1}\ [\eta_{s'},\ H_0]\ \sigma(H_0 - \lambda\ \mp\ i\,\sigma)^{-1}\ \eta_{-s'}\ v_1 = 0, \end{aligned} \tag{3.81}$$

where we used Theorem 2.5, the spectral theorem, that H_0 has no non zero eigenvalues, and the fact that for $\sigma > 0$, $(H_0 - \lambda \mp i\,\sigma)^{-1} \in \mathcal{B}(L^2_{s'}(\mathbf{R}^3),\ L^2_{s'}(\mathbf{R}^3))$, $\quad s' \in \mathbf{R}$ (this follows as in the proof of Lemma 4.1 in Chapter 2 and duality).

Then by (3.75)

$$v_1 = \hat{R}_{0,\pm}(\lambda \pm i\,0)\ \chi(-\Delta)\ f(H_0) q_1\ \hat{u}. \tag{3.82}$$

By taking the Fourier transform in the x variables we obtain for almost every $k \in \mathbf{R}^2$,

$$\tilde{v}_1(z,k) = R_k(\lambda \pm i\,0)\ \chi(k^2)\ f(H_0(k))(q_1\ \hat{u})^{\sim}, \tag{3.83}$$

where $\tilde{v}_1(z,k)$, and $(q\ \hat{u})^{\sim}(z,k)$ are respectively the Fourier transform in x of $v_1(x,z)$ and $q_1(x,z)\hat{u}(x,z)$. As (3.83) holds with both the $+$ and $-$ signs it follows from (3.38) that

$$\chi(k^2) B_k(\lambda)(1+z^2)^{s/2}\ f(H(k))\ (q_1\ \hat{u})^{\sim} = 0. \tag{3.84}$$

Let us denote

$$w(z,k) = (1+z^2)^{-s/2}\ \tilde{v}_1(z,k), \tag{3.85}$$

with $s = \frac{1+\delta}{2}$. Then by (3.38) and (3.39)

$$w(z,k) = w_\lambda(z,k) + w_{\hat{\lambda}}(z,k), \tag{3.86}$$

where

$$w_\lambda(z,k) =$$

$$\chi(k^2) J_k\left(\frac{1}{\rho-\lambda}(B_k(\rho) - B_k(\lambda))\ (1+z^2)^{s/2}\ f(H(k))\ (q_1\ \hat{u})^{\sim}\right), \tag{3.87}$$

$$w_{\hat{\lambda}}(z,k) = \chi(k^2)\ (1+z^2)^{-s/2}\ E_k(I^{\sim}_{\lambda})\ R_k(\lambda)\ f(H(k))(q_1\ \hat{u})^{\sim}. \tag{3.88}$$

Note that as in Lemma 3.4 we prove that $f(H(k)) \in \mathcal{B}(L^{2,6}_s(\mathbf{R}),\ L^{2,6}_s(\mathbf{R}))$, for $-1 < s < 1$.

It follows from (3.75) and (3.80) that

$$(H^2(k) - \lambda^2)\ \tilde{v}_2(z,k) = (1 - \chi(k^2))\ (H(k) + \lambda)\ f(H(k))\ (q_1\ \hat{u})^{\sim}, \tag{3.89}$$

where $\tilde{v}_2(z,k)$ is the Fourier transform in x of $v_2(x,z)$.

Let us denote by P_E and P_M the following operators from $\mathbf{C}^6$ onto $\mathbf{C}^3$

$$P_E\, v = (v_1,\ v_2,\ v_3), \tag{3.90}$$

$$P_M\, v = (v_4,\ v_5,\ v_6), \tag{3.91}$$

for

$$v = (v_1,\ v_2,\ v_3,\ v_4,\ v_5,\ v_6) \in \mathbf{C}^6. \tag{3.92}$$

We denote by

$$O_E = \frac{1}{c_0^2}\, \mathcal{F}^{-1}\, (I - \chi(k^2))\, V_E(|k|)\, P_E U(k)\, f(H(k))\, \mathcal{F}, \tag{3.93}$$

and

$$O_M = \frac{1}{c_0^2}\, \mathcal{F}^{-1}\, (I - \chi(k^2))\, V_M(|k|)\, P_M U(k)\, f(H(k))\, \mathcal{F}. \tag{3.94}$$

Note that since $\chi(k^2) = 1$ in a neighborhood of $k = 0$, O_E and O_M are bounded operators from $L_s^{2,6}(\mathbf{R}^3)$ respectively into $P_E\ L_s^{2,6}(\mathbf{R}^3)$ and $P_M\ L_s^{2,6}(\mathbf{R}^3)$, for $-1 < s < 1$.

We define

$$\varphi_E(x,z) = \mathcal{F}^{-1}\, V_E(|k|)\, P_E U(k)\, (1 - \chi(k^2))\, \mathcal{F}\, u_2, \tag{3.95}$$

$$\varphi_M(x,z) = \mathcal{F}^{-1}\, V_M(|k|)\, P_M U(k)\, (1 - \chi(k^2))\, \mathcal{F}\, u_2, \tag{3.96}$$

where $V_E(|k|)$, $V_M(|k|)$ are defined as in section 1 (see (1.56)).

We have that

$$v_2 = G[\varphi_E \oplus \varphi_M], \tag{3.97}$$

where

$$G = \mathcal{F}^{-1}\, U^*(k)(V_E^*(|k|)\ \oplus\ V_M^*(|k|))\ g(|k|^2)\ \mathcal{F}, \tag{3.98}$$

where $g(\rho) \in C^\infty(\mathbf{R})$, $g(\rho) = 0$ for ρ in a neighborhood of $\rho = 0$, and $g(\rho^2) = 1$ for ρ in the support of $(1 - \chi(\rho^2))$. Moreover

$$V_E^*(|k|) \begin{bmatrix} \psi_1 \\ \psi_2 \end{bmatrix} = \begin{bmatrix} -\psi_2 \\ \frac{1}{\mu_0}\frac{d}{dz}\psi_1 \\ \frac{|k|}{i\,\mu_0}\,\psi_1 \end{bmatrix}, \tag{3.99}$$

$$V_M^*(|k|) \begin{bmatrix} \psi_1 \\ \psi_2 \end{bmatrix} = \begin{bmatrix} -\psi_2 \\ \frac{1}{\epsilon_0}\frac{d}{dz}\psi_1 \\ \frac{|k|}{i\,\epsilon_0}\,\psi_1 \end{bmatrix}. \tag{3.100}$$

Note that $G \in \mathcal{B}\left(\bigoplus_{i=1}^{4} L^2_s(\mathbf{R}^3),\ L^{2,6}_s(\mathbf{R}^3)\right)$, for $-1 < s < 1$.

It follows from (1.28), (1.57), (1.58), and (3.89) that

$$\left(B_E - \lambda^2\ c_+^{-2}\right)\varphi_E = O_E\ q_1\ \hat{u}, \tag{3.101}$$

$$\left(B_M - \lambda^2\ c_+^{-2}\right)\varphi_M = O_M\ q_1\ \hat{u}, \tag{3.102}$$

where B_E and B_M are the operators (see Appendix 2 for details)

$$B_E = -\mu_0\ \frac{d}{dz}\ \frac{1}{\mu_0}\ \frac{d}{dz} - \frac{\partial^2}{\partial^2 x_1} - \frac{\partial^2}{\partial^2 x_2} + q(z), \tag{3.103}$$

$$B_M = -\epsilon_0\ \frac{d}{dz}\ \frac{1}{\epsilon_0}\ \frac{d}{dz} - \frac{\partial^2}{\partial^2 x_1} - \frac{\partial^2}{\partial^2 x_2} + q(z), \tag{3.104}$$

where

$$q(z) = -\lambda^2\ c_0^{-2}(z) + \lambda^2\ c_+^{-2}. \tag{3.105}$$

In what follows we denote by φ_D, O_D, and B_D either φ_E, O_E, and B_E, or φ_M, O_M, and B_M.

As in the proof of formulas (6.77) and (6.78) in Section 6 of Chapter 2, we have that

$$\varphi_D = R_{B_D}(\lambda^2\ c_+^{-2} \pm i\ 0) O_D\ q_1\ \hat{u}, \tag{3.106}$$

where we used the limiting absorption principle for B_D (see formula (1.70) in Appendix 2), and the fact that $\psi_D \in (L^2_{-s'}(\mathbf{R}^3) \oplus L^2_{-s'}(\mathbf{R}^3))$, $0 < s' < 1/2$.

Since (3.106) holds both with the + and - signs it follows from formula (1.71) in Appendix 2 that

$$C_D^{\pm}(\tilde{\lambda})\eta_s\ O_D\ q_1\ \hat{u} = 0, \tag{3.107}$$

$$C_D^{0}(\tilde{\lambda})\eta_s\ O_D\ q_1\ \hat{u} = 0, \tag{3.108}$$

$$C^{j}(\tilde{\lambda}_j)\eta_s\ O_D\ q_1\ \hat{u} = 0, \tag{3.109}$$

for $j = 1, 2, 3, \cdots Q$, where $C_E^{\pm}$, C_E^{j}, and $C_M^{\pm}$, C_M^{j}, are respectively the operators $C^{\pm}$ and C^{j} (see formulas (1.66) and (1.67) in Appendix 2) for B_E and B_M, $\tilde{\lambda} = (\lambda^2\ c_+^{-2} - q_-)^{1/2}$, $\tilde{\lambda}_j = (\lambda^2\ c_+^{-2} - \lambda_j)^{1/2}$, $j = 1, 2, 3, \cdots$, and

$$q_- = -\lambda^2\ c_-^{-2} + \lambda^2\ c_+^{-2}. \tag{3.110}$$

We denote

$$\psi_D = \eta_{-s}(x, z)\ \varphi_D(x, z), \tag{3.111}$$

where $s = \frac{1+\delta}{2}$.

By (3.106) and formula (1.71) in Appendix 2

$$\psi_D = \psi_{\lambda,D} + \psi_{\hat{\lambda},D}, \tag{3.112}$$

where

$\psi_{\lambda,D} =$

$$M_{D,0}\left(\frac{1}{\rho^2 + q_- - \lambda^2\, c_+^{-2}}\left(C_D^+(\rho)\,\oplus\, C_D^-(\rho)\,\oplus\, C_D^0(\rho)\right)\eta_s\, O_D\, q_1\, \hat{u}\right) +$$

$$+\sum_{j=1}^{Q} M_{D,j}\left(\frac{1}{\rho^2 + \lambda_j - \lambda^2\, c_+^{-2}}\, C_D^j(\rho)\,\eta_s\, O_D\, q_1\, \hat{u}\right), \tag{3.113}$$

where $M_{E,j},\ M_{M,j},\ j = 0,1,2,\cdots Q$, are respectively the operators M_j (see formulas (1.73) and (1.76) in Appendix 2) for B_E and B_M, and

$$\psi_{\hat{\lambda},D} = \eta_{-s}\, E_{B_D}([a,b]^{\sim})\, R_{B_D}(\lambda^2\, c_+^{-2})\, O_D\, q_1\, \hat{u}. \tag{3.114}$$

It follows from (3.88) and (3.114) that

$$\left\| w_{\hat{\lambda}}(k,z)\right\|_{L_s^{2,6}(\mathbf{R})} \leq C\,\chi(k^2)\,\left\|(q_1\,\hat{u})^{\sim}\right\|_{\mathcal{L}}, \tag{3.115}$$

$$\left\|\psi_{\hat{\lambda},D}\right\|_{L_s^2(\mathbf{R}^3)\,\oplus\,L_s^2(\mathbf{R}^3)} \leq C\,\left\|q_1\,\hat{u}\right\|_{\mathcal{H}_0}. \tag{3.116}$$

Suppose that $\hat{u} \in L^{2,6}_{-s'}(\mathbf{R}^3)$, for some $0 < s' < s$. Then by (3.87)

$$\left\| w_\lambda(z,k)\right\|_{L_{rs}^{2,6}(\mathbf{R})} \leq C\,\chi(k^2)\,\left\|\hat{u}\right\|_{L_{-s'}^{2,6}(\mathbf{R}^3)}, \tag{3.117}$$

for any $r < 2\,(2s - s' - 1/2)$, $r \leq 1$. Note that it follows from the explicit formulas for $\phi_\pm(z,k,\lambda)$ that the constant C in (3.117) is uniform for all k in the support of $\chi(k^2)$.

Furthermore it follows from (3.113) that

$$\left\|\psi_{\lambda,D}\right\|_{L_{rs}^2(\mathbf{R}^3)\,\oplus\,L_{rs}^2(\mathbf{R}^3)} \leq C\,\left\|\hat{u}\right\|_{L_{-s'}^{2,6}(\mathbf{R}^3)}, \tag{3.118}$$

for all $r < 2\,(2s - s' - 1/2)$, $r \leq 1$.

But then by (3.77), and (3.115) to (3.118)

$$\hat{u} \in L^{2,6}_{rs-s}(\mathbf{R}^3), \tag{3.119}$$

where we also used (3.97) and (3.111). Iterating this argument we prove that $\hat{u} \in L^{2,6}(\mathbf{R}^3)$, and by (3.71) that λ is an eigenvalue of H.

Q. E. D.

Lemma 3.6

Suppose that (3.24) and (3.25) are true. Then the non zero eigenvalues of H have finite multiplicity and can accumulate only at zero and $\pm\infty$.

Proof: Suppose that the lemma is not true. Then there will be a sequence σ_i of eigenvalues contained in a compact set of $\mathbf{R}\setminus 0$, with an associated orthonormal sequence of eigenfunctions u_i in $\mathcal{H}$. Then by (3.17), and (3.23)

$$Q^*_{0,-}(\sigma_i)u_i = 0, \tag{3.120}$$

and by Remark 3.1

$$u_i = -Q^*(\lambda_0)\ K^*_-(\sigma_i)\ Q^*_0(\lambda_0)u_i. \tag{3.121}$$

As $K^*_-(\sigma_i)$ is a norm continuous sequence of compact operators on $L^{2,6}_{-s}(\mathbf{R}^3)$, $s=\frac{1+\delta}{2}$, there exist subsequences that we also denote by σ_i, u_i, and $\sigma_\infty\in\mathbf{R}\setminus 0$, $u_\infty\in L^{2,6}_{-s}(\mathbf{R}^3)$, such that $\lim\limits_{i\to\infty}\sigma_i=\sigma_\infty$, and $\lim\limits_{i\to\infty}u_i=u_\infty$, in the strong topology in $L^{2,6}_{-s}(\mathbf{R}^3)$. As $\|u_i\|_{\mathcal{H}}=1$, we can as well assume that $u_\infty\in\mathcal{H}$.

By (3.120) and (3.23)

$$u_i = \sigma_i\ R_{0,+}(\sigma_i)\ (e^{-2}-I)u_i, \tag{3.122}$$

for $i=1,2,3,\cdots$, and $i=\infty$.

As in (3.72), and (3.73) we define

$$u_{i,1} = (I-f(H_0))u_i, \tag{3.123}$$

$$u_{i,2} = (I-f(H_0))u_i, \tag{3.124}$$

$i=1,2,3,\cdots,\infty$, where $f(\rho)\in C^\infty_0(\mathbf{R}\setminus 0)$ satisfies $f(\rho)\equiv 1$ in a neighborhood of the σ_i. Then

$$\lim_{i\to\infty}\left\|u_{i,1}-u_{\infty,1}\right\|_{\mathcal{H}} = \lim_{i\to\infty}\left\|\Big(I-f(H_0)\Big)\Big(\sigma_i\ R_0(\sigma_i)\right.$$

$$\left.(e^{-2}-I)u_i-\sigma_\infty\ R_0(\sigma_\infty)\ (e^{-2}-I)u_\infty\Big)\right\|_{\mathcal{H}} = 0. \tag{3.125}$$

As in (3.79) and (3.80) we define

$$v_{i,1} = \chi(-\Delta)u_{i,2}, \tag{3.126}$$

$$v_{i,2} = (1-\chi(-\Delta))u_{i,2}, \tag{3.127}$$

and as in (3.83) for i large enough

$$\tilde{v}_{i,1}(z,k) = R_k(\sigma_i \pm i\,0)\ \chi(k^2)\ f(H_0(k))\ ((I-e^{-2})u_i)^{\sim}, \tag{3.128}$$

where $\tilde{v}_{i,1}(z,k)$ and $((I-e^{-2})u_i)^{\sim}(z,k)$ are the Fourier transform in x of $v_{i,1}(x,z)$ and $((I-e^{-2})u_i)(x,z)$. But as in (3.86) to (3.88)

$$\tilde{v}_{i,1}(z,k) = \tilde{v}_{i,1,\sigma_i}(z,k) + \tilde{v}_{i,1,\hat{\sigma}_i}(z,k), \tag{3.129}$$

where

$$\tilde{v}_{i,1,\sigma_i} = (1+z^2)^{s/2}\ \chi(k^2)\ J_k\left(\frac{1}{\rho-\sigma_i}\ (B_k(\rho)-B_k(\sigma_i))\ (1+z^2)^{s/2} \cdot f(H(k))\ ((e^{-2}-I)u_i)^{\sim}\right), \tag{3.130}$$

$$\tilde{v}_{i,1,\hat{\sigma}_i} = \chi(k^2)\ E_k(I^{\sim}_{\sigma_\infty})\ R_k(\sigma_i)\ f(H(k))\ ((e^{-2}-I)u_i)^{\sim}\ . \tag{3.131}$$

Moreover,

$$\lim_{i\to\infty} \left\|\tilde{v}_{i,1,\sigma_i} - \tilde{v}_{\infty,1,\sigma_\infty}\right\|_{L^{2,6}(\mathbf{R}^3)} \leq C \lim_{i\to\infty} \int dk\ \chi(k^2)\ \Big\|E_k(I^{\sim}_{\sigma_\infty}) \cdot f(H(k))\Big(R_k(\sigma_i)\ ((e^{-2}-I)u_i)^{\sim} - R_k(\sigma_\infty)((e^{-2}-I)u_\infty)^{\sim}\Big)\Big\|_{\mathcal{H}_0} = 0. \tag{3.132}$$

Moreover, we have that

$$\tilde{v}_{i,1,\sigma_i} - \tilde{v}_{\infty,1,\sigma_\infty} = (1+z^2)^{s/2}\ \chi(k^2)\ J_k(g_i), \tag{3.133}$$

where

$$g_i(\rho) = \frac{1}{\rho-\sigma_i}\ \Big[(B_k(\rho)-B_k(\sigma_i))\ (1+z^2)^{s/2}\ f(H(k)) \cdot \sigma_i((e^{-2}-I)u_i)^{\sim} - (B_k(\rho)-B_k(\sigma_\infty))\ (1+z^2)^{s/2}\ f(H(k)) \cdot \sigma_\infty((e^{-2}-I)u_\infty)^{\sim}\Big]. \tag{3.134}$$

Note that for $\rho \neq \sigma_\infty$ and ρ^2 in the support of $\chi(\rho^2)$

$$\lim_{i\to\infty} g_i(\rho) = 0, \tag{3.135}$$

$$|g_i(\rho)| \leq C \left(\frac{1}{|\rho - \sigma_i|^{1/2-\Delta}} + \frac{1}{|\rho - \sigma_\infty|^{1/2-\Delta}} \right), \tag{3.136}$$

for some $\Delta > 0$. Then by dominated convergence (see Theorem 16, Section 4 of Chapter 4 of Royden 1968)

$$\begin{aligned} &\lim_{i\to\infty} \left\| \tilde{v}_{i,1,\sigma_i} - \tilde{v}_{\infty,1,\sigma_\infty} \right\|^2_{L^{2,6}(\mathbf{R}^3)} \leq \\ &\leq C \lim_{i\to\infty} \int dk\ \chi(k^2) \int_{I_{\sigma_\infty}} d\rho\ |g_i(\rho)|^2 = 0. \end{aligned} \tag{3.137}$$

By (3.126), (3.129), (3.132), and (3.137)

$$\lim_{i\to\infty} \left\| v_{i,1} - v_{\infty,1} \right\|_{\mathcal{H}} = 0. \tag{3.138}$$

We consider now $v_{i,2}$, defining as in (3.95) and (3.96)

$$\varphi_{i,E}(x,z) = \mathcal{F}^{-1}\ V_E(|k|)\ P_E\ U(|k|)\ (1 - \chi(k^2))\ \mathcal{F} u_{i,2}, \tag{3.139}$$

$$\varphi_{i,M}(x,z) = \mathcal{F}^{-1}\ V_M(|k|)\ P_M\ U(|k|)\ (1 - \chi(k^2))\ \mathcal{F} u_{i,2}, \tag{3.140}$$

for $i = 1, 2, \cdots, \infty$.

Then as in (3.97)

$$v_{i,2} = G[\varphi_{i,E}\ \oplus\ \varphi_{i,M}]. \tag{3.141}$$

As in (3.112) for $D = E$, or $D = M$

$$\varphi_{i,D} = \varphi_{i,\sigma_i,D} + \varphi_{i,\hat{\sigma}_i,D}, \tag{3.142}$$

where

$$\begin{aligned} &\varphi_{i,\sigma_i,D} = \\ &\eta_s\ M_{D,0} \left(\frac{1}{\rho^2 + q_- - \sigma_i^2\ c_+^{-2}}\ (C_D^+(\rho) \oplus C_D^-(\rho) \oplus C_D^0(\rho)) \eta_s O_D(e^{-2} - I) u_i \right) \\ &+\ \eta_s \sum_{j=1}^{Q} M_{D,j} \left(\frac{1}{\rho^2 + \lambda_j - \sigma_i^2\ c_+^{-2}}\ C_D^j(\rho)\ \eta_s O_D(e^{-2} - I) u_i \right), \end{aligned} \tag{3.143}$$

$$\varphi_{i,\hat{\sigma}_i,D} = E_{B_D}([a,b]^\sim)\ R_{B_D}(\sigma_i^2\ c_+^{-2})\ O_D(e^{-2} - I) u_i. \tag{3.144}$$

We have that

$$\lim_{i\to\infty} \left\| \varphi_{i,\hat{\sigma}_i,D} - \varphi_{\infty,\hat{\sigma}_\infty,D} \right\|_{L^2(\mathbf{R}^3)\oplus L^2(\mathbf{R}^3)} = 0. \tag{3.145}$$

Moreover,

$$\varphi_{i,\sigma_i,D} - \varphi_{\infty,\sigma_\infty,D} = \eta_s \sum_{j=0}^{Q} M_{D,j}(g_{i,j,D}), \tag{3.146}$$

where

$$g_{i,0,D} = \frac{\sigma_i}{\rho^2 + q_- - \sigma_i^2\ c_+^{-2}} \Big(C_D^+(\rho) - C_D^+(\tilde{\sigma}_i) \oplus C_D^-(\rho) - C_D^-(\tilde{\sigma}_i)$$

$$\oplus\ C_D^0(\rho) - C_D^0(\tilde{\sigma}_i) \Big) \eta_s O_D(e^{-2} - I) u_i - \frac{\sigma_\infty}{\rho^2 + q_- - \sigma_\infty^2\ c_+^{-2}}$$

$$\cdot \Big(C_D^+(\rho) - C_D^+(\tilde{\sigma}_\infty) \oplus\ C_D^-(\rho) - C_D^-(\tilde{\sigma}_\infty) \oplus C_D^0(\rho)$$

$$- C_D^0(\tilde{\sigma}_\infty) \Big) \eta_s O_D(e^{-2} - I) u_\infty, \tag{3.147}$$

and for $j = 1, 2, 3, \cdots, Q$

$$g_{i,j,D} = \frac{\sigma_i}{\rho^2 + \lambda_j - \sigma_i^2\ c_+^{-2}} \Big(C_D^j(\rho) - C_D^j(\tilde{\sigma}_{i,j}) \Big) \eta_s O_D(e^{-2} - I) u_i -$$

$$- \frac{\sigma_\infty}{\rho^2 + \lambda_j - \sigma_\infty^2\ c_+^{-2}} \Big(C_D^j(\rho) - C_D^j(\tilde{\sigma}_{\infty,j}) \Big) \eta_s O_D(e^{-2} - I) u_\infty, \tag{3.148}$$

where $\tilde{\sigma}_i = (\sigma_i^2\ c_+^{-2} - q_-)^{1/2}$, $\tilde{\sigma}_{i,j} = (\sigma_i^2\ c_+^{-2} - \lambda_j)^{1/2}$.

We have that for $\rho \neq \tilde{\sigma}_\infty$, $\rho \neq \tilde{\sigma}_{\infty,j}$

$$\lim_{i\to\infty} g_{i,j,D} = 0, \tag{3.149}$$

and for some $\Delta > 0$

$$|g_{i,0,D}(\rho)| \leq C \left(\frac{1}{(\rho^2 - \tilde{\sigma}_i^2)^{1/2-\Delta}} + \frac{1}{(\rho^2 - \tilde{\sigma}_\infty^2)^{1/2-\Delta}} \right), \tag{3.150}$$

$$|g_{i,j,D}(\rho)| \leq C \left(\frac{1}{(\rho^2 - \tilde{\sigma}_{i,j}^2)^{1/2-\Delta}} + \frac{1}{(\rho^2 - \tilde{\sigma}_{\infty,j}^2)^{1/2-\Delta}} \right), \tag{3.151}$$

for $j = 1, 2, 3, \cdots, Q$.

Then by (3.146) and dominated convergence

$$\lim_{i\to\infty} \left\|\varphi_{i,\sigma_i,D} - \varphi_{\infty,\sigma_\infty,D}\right\|_{L^2(\mathbf{R}^3)\,\oplus\, L^2(\mathbf{R}^3)} = 0. \tag{3.152}$$

It follows from (3.141), (3.142), (3.145), and (3.152) that

$$\lim_{i\to\infty} \left\|v_{i,2} - v_{\infty,2}\right\|_{\mathcal{H}} = 0, \tag{3.153}$$

and by (3.126), (3.127), and (3.138) that

$$\lim_{i\to\infty} \left\|u_i - u_\infty\right\|_{\mathcal{H}} = 0, \tag{3.154}$$

but this is a contradiction since the u_i are an orthonormal sequence in $\mathcal{H}$.

Q. E. D.

Theorem 3.7

Assume that (3.4), (3.24), and (3.25) are satisfied. Then the non zero eigenvalues of H have finite multiplicity and can only accumulate at 0 and $\pm\,\infty$. H has no singular continuous spectrum.

For every $\lambda \in \mathbf{R} \setminus \sigma_p(H)$ the limits

$$R(\lambda \pm i\,0) \doteq \lim_{\sigma\downarrow 0} R(\lambda \pm i\,\sigma) \tag{3.155}$$

exist in the uniform operator topology on $\mathcal{B}(L^{2,6}_s(\mathbf{R}^3),\ L^{2,6}_{-s}(\mathbf{R}^3))$, $s = \frac{1+\delta}{2}$, uniformly for λ in compact sets of $\mathbf{R} \setminus \sigma_p(H)$. Moreover the functions

$$R_\pm(\lambda) = \begin{cases} R(\lambda), & Im\ \lambda \neq 0, \\ R(\lambda \pm i\,0), & \lambda \in \mathbf{R} \setminus \sigma_p(H), \end{cases} \tag{3.156}$$

defined for $\lambda \in \mathbf{C}^\pm \cup (\mathbf{R} \setminus \sigma_p(H))$ are analytic for $Im\ \lambda \neq 0$ and locally Hölder continuous for $\lambda \in \mathbf{R} \setminus \sigma_p(H)$ with exponent γ that satisfies $\gamma < 1$, $\gamma < (s - 1/2)$, if $\lambda \neq \pm\, c_+\ \rho_j^E$, $\lambda \neq \pm\, c_+\ \rho_j^M$, $j = 1, 2, 3, \cdots$, and $\gamma < 1/2$, $\gamma < (s - 1/2)$ if $\lambda = \pm\, c_+\ \rho_j^E$, or $\lambda = \pm\, c_+\ \rho_j^M$, for some $j = 1, 2, 3, \cdots$.

Proof: We define for $\lambda \in \mathbf{R} \setminus \sigma_p(H)$ (see (3.26))

$$R(\lambda \pm i\, 0) = R_{0,\pm}(\lambda)\; Q_{0,\pm}^{-1}(\lambda)\; e^{-2}. \tag{3.157}$$

Then the theorem follows by Theorem 2.5, and by Lemmas 3.5 and 3.6.

Q. E. D.

Remark 3.8

It follows from Remark 3.1 and (3.157) that

$$R(\lambda \pm i\, 0) = R_{0,\pm}(\lambda)\; (I + K_{\pm}(\lambda))^{-1}\; Q(\lambda_0)\; e^{-2}. \tag{3.158}$$

We study below the continuity of $R_{\pm}(\lambda)$ on the perturbation. We first define

Definition 3.9

Let $\epsilon(x,z)$ and $\mu(x,z)$ be real valued measurable functions on $\mathbf{R}^3$ that satisfy (3.4), (3.24), and (3.25). A neighborhood, of ϵ, μ, $O_{\epsilon,\mu}$, is defined as the set of all real valued measurable functions on $\mathbf{R}^3$, $\tilde{\epsilon}(x,z)$, $\tilde{\mu}(x,z)$ that satisfy (3.4), (3.24), and (3.25) with the same constants c_m, c_M and δ as $\epsilon(x,z)$ and $\mu(x,z)$, and such that for some fixed $\Delta > 0$

$$\left\| \eta_{1+\delta}(|\epsilon - \tilde{\epsilon}| + |\mu - \tilde{\mu}|) \right\|_{\infty} \leq \Delta. \tag{3.159}$$

We will denote by $H_{\epsilon,\mu}$ the operator (3.6) with coefficients ϵ, μ, and similarly by $H_{\tilde{\epsilon},\tilde{\mu}}$ the operator (3.6) with coefficients $\tilde{\epsilon}$, $\tilde{\mu}$, instead of ϵ and μ. We denote respectively by $R_{\pm,\epsilon,\mu}(\lambda)$, and $R_{\pm,\tilde{\epsilon},\tilde{\mu}}(\lambda)$ the extended resolvents (3.156) of $H_{\epsilon,\mu}$, and $H_{\tilde{\epsilon},\tilde{\mu}}$.

Corollary 3.10

The extended resolvents $R_{\pm,\epsilon,\mu}(\lambda)$ are locally Lipschitz continuous on $\epsilon(x,z)$ and $\mu(x,z)$ uniformly for λ in compact sets of $\mathbf{R} \setminus \sigma_p(H_{\epsilon,\mu})$. That is to say for each compact set $K \subset \mathbf{R} \setminus \sigma_p(H_{\epsilon,\mu})$ there is a neighborhood of ϵ, μ, $O_{\epsilon,\mu}$, such that for all $\tilde{\epsilon}(x,z), \tilde{\mu}(x,z) \in O_{\epsilon,\mu}$, $K \subset \mathbf{R} \setminus \sigma_p(H_{\tilde{\epsilon},\tilde{\mu}})$ and

$$\left\| R_{\pm,\epsilon,\mu}(\lambda) - R_{\pm,\tilde{\epsilon},\tilde{\mu}}(\lambda) \right\|_{\mathcal{B}\left(L^{2,6}_s(\mathbf{R}^3),\ L^{2,6}_{-s}(\mathbf{R}^3)\right)}$$

$$\leq C \left\| (1+|x|+|z|)^{1+\delta} \, (|\epsilon-\tilde{\epsilon}| + |\mu-\tilde{\mu}|) \right\|_\infty . \quad (3.160)$$

Proof: By the stability of bounded invertibility theorem (see Theorem 1.16, Chapter IV, Section 1 of Kato 1976) (3.23), and Lemma 3.5, there is a neighborhood, $O_{\epsilon,\mu}$ of ϵ, μ such that for all $\tilde{\epsilon}, \tilde{\mu} \in O_{\epsilon,\mu}$, $K \subset \mathbf{R} \setminus \sigma_p(H_{\tilde{\epsilon},\tilde{\mu}})$, and that (3.160) holds (see (3.157))

Q. E. D.

§4. The Generalized Fourier Maps and Electromagnetic Scattering Theory

As in the case of acoustic waves we use in this section the results on the limiting absorption principle to construct the generalized Fourier maps for the perturbed electromagnetic propagator, H, and to study the electromagnetic scattering theory.

As in Section 7 of Chapter 1, it follows from (2.62) and Stone's formula that for every $\lambda \in \mathbf{R} \setminus 0$, $u, v \in \mathcal{H}_0$, $s > 1/2$

$$\frac{d}{d\lambda} (E_{H_0}(\lambda)\eta_{-s}u,\ \eta_{-s}v)_{\mathcal{H}_0} = (B(\lambda)u,\ B(\lambda)v)_{\mathcal{H}_0}. \quad (4.1)$$

Then by (3.157) and Stone's formula for every $\lambda \in \mathbf{R} \setminus \sigma_p(H)$

$$\frac{d}{d\lambda} \left(E_H(\lambda)\eta_{-s}u,\ \eta_{-s}v\right)_{\mathcal{H}} = \Big(B(\lambda)\eta_s Q_\pm(\lambda)e^{-2}\eta_{-s}u,$$

$$\cdot \quad B(\lambda)\eta_s Q_\pm(\lambda)e^{-2}\eta_{-s}v\Big)_{\hat{\mathcal{H}}(\lambda)}, \quad (4.2)$$

where

$$Q_\pm(\lambda) = (H_0 - \lambda)R_\pm(\lambda)e^2 = I + \lambda(e^{-2} - I)R_\pm(\lambda)e^2. \quad (4.3)$$

We define the generalized Fourier maps, $F^{\pm}$, for the perturbed electromagnetic propagator, first on vectors of the form

$$\psi = \sum_{i=1}^{N} E_H(I_i)\, \eta_{-s}\, \psi_i, \tag{4.4}$$

where $I_i \subset \mathbf{R} \setminus \sigma_p(H)$, $I_i \cap I_j = \emptyset$, $i \neq j$, $s > 1/2$, and $\psi_j \in \mathcal{H}$, as

$$F^{\pm}\, \psi = \sum_{i=1}^{N} \chi_{I_i}(\lambda) B(\lambda) \eta_s Q_{\pm}(\lambda) e^{-2} \eta_{-s} \psi_i. \tag{4.5}$$

By (4.2)

$$\left\| F^{\pm}\, \psi \right\|_{\hat{\mathcal{H}}} = \left\| \psi \right\|_{\mathcal{H}}. \tag{4.6}$$

We extend the $F^{\pm}$ by continuity as isometric operators from $\mathcal{H}_{ac}(H)$ into $\hat{\mathcal{H}}$. We moreover extend the $F^{\pm}$ as partially isometric operators from $\mathcal{H}$ onto $\hat{\mathcal{H}}$ by defining $F^{\pm} = 0$ on $\mathcal{H}_{pp}(H)$.

Lemma 4.1

Suppose that (3.4), (3.24), and (3.25) are satisfied. Then the generalized Fourier maps, $F^{\pm}$, have the following properties:

1.

$$(F^{\pm}) = \mathcal{H}_{pp}(H). \tag{4.7}$$

2. The $F^{\pm}$ are unitary operators from $\mathcal{H}_{ac}(H)$ onto $\hat{\mathcal{H}}$, and furthermore,

$$H\, P_{ac}(H) = F^{\pm *}\, \lambda\, F^{\pm}. \tag{4.8}$$

Moreover, $\sigma(H) = \mathbf{R}$.

Proof: Property 1 is immediate from the definition of the $F^{\pm}$. The proof of the rest of the Lemma follows as in Lemma 7.1, in Chapter 2. The fact that $\sigma(H) = \mathbf{R}$ is immediate from (4.8).

Q. E. D.

Let us define the following operator of identification from $\mathcal{H}_0$ onto $\mathcal{H}$ as

$$(Ju)(x,z) = u(x,z). \tag{4.9}$$

The electromagnetic wave operators are defined as

$$W^{\pm} = s - \lim_{t \to \pm\infty} e^{itH}\, J\, e^{-itH_0}\, P^0, \tag{4.10}$$

provided that the strong limits exist in $\mathcal{H}$, and where P^0 denotes the projector onto the orthogonal complement to the kernel of H_0. We define the electromagnetic scattering operator as

$$S = W^{+*}\, W^{-}. \tag{4.11}$$

Definition 4.2

We say that a real valued measurable function, $f(\lambda)$, defined on $\mathbf{R}$ is admissible if for all $\varphi(\lambda) \in L^2(\mathbf{R})$

$$\lim_{t\to\infty} \int_0^\infty \Big| \int_{\mathbf{R}} e^{-itf(\lambda)-is\lambda}\, \varphi(\lambda) d\lambda \Big|^2 \, ds = 0. \tag{4.12}$$

Note the difference with Definition 7.2 in Chapter 2. This is necessary because the spectrum of H is all of $\mathbf{R}$. Clearly $f(\lambda) = \lambda$ is admissible, and as in the case of Definition 7.2 in Chapter 2, $f(\lambda)$ is admissible if the real axis can be divided into a finite number of subintervals in such a way that in each open subinterval $f(\lambda)$ is differentiable with $\frac{d}{d\lambda} f(\lambda)$ continuous locally of bounded variation, and positive.

Lemma 4.3

If (3.4), (3.24), and (3.25) are satisfied the wave operators $W^{\pm}$ exist and are unitary from $P^0\mathcal{H}_0$ onto $\mathcal{H}_{ac}(H)$. The scattering operator, S, is unitary and the following stationary formulas for the wave operators are valid

$$W^{\pm} = F^{\pm *}\, F. \tag{4.13}$$

Finally the invariance principle holds: for any admissible function $f(\lambda)$

$$W^{\pm} = s - \lim_{t\to\pm\,\infty} e^{itf(H)}\, J\, e^{-itf(H_0)}\, P^0. \tag{4.14}$$

Proof: The lemma is proven as Lemma 7.3 in Chapter 2.

Q. E. D.

We define

$$\hat{S} = F\ S\ F^* = F^+\ F^{-*}. \tag{4.15}$$

Lemma 4.4

We assume that (3.4), (3.24), and (3.25) are true. For $\lambda \in \mathbf{R} \setminus \sigma_p(H)$ we define

$$S(\lambda) = I - 2\ \pi\ i\ \lambda\ B(\lambda)\eta_s Q_+(\lambda)\ (I - e^{-2})\eta_s B^*(\lambda). \tag{4.16}$$

$S(\lambda)$ is a unitary operator on $\hat{\mathcal{H}}(\lambda)$ with inverse given by

$$S^{-1}(\lambda) = I + 2\ \pi\ i\ \lambda\ B(\lambda)\eta_s Q_-(\lambda)\ (I - e^{-2})\eta_s B^*(\lambda). \tag{4.17}$$

Furthermore, $\hat{S}$ and $\hat{S}^{-1}$ are the direct integrals of $S(\lambda)$ and $S^{-1}(\lambda)$

$$\hat{S} = \oplus \int_{\mathbf{R}} S(\lambda) d\lambda, \tag{4.18}$$

$$\hat{S}^{-1} = \oplus \int_{\mathbf{R}} S^{-1}(\lambda) d\lambda. \tag{4.19}$$

That is to say for any

$$u(\lambda) \in \hat{\mathcal{H}} = \oplus \int_{\mathbf{R}} \hat{\mathcal{H}}(\lambda) d\lambda, \tag{4.20}$$

$$(\hat{S}\ \varphi)(\lambda) = S(\lambda)\varphi(\lambda), \tag{4.21}$$

$$(\hat{S}^{-1}\ \varphi)(\lambda) = S^{-1}(\lambda)\varphi(\lambda). \tag{4.22}$$

$S(\lambda) - I$ is a compact operator on $\hat{\mathcal{H}}(\lambda)$, and the function $\lambda \to S(\lambda)$ from $\mathbf{R} \setminus \sigma_p(H)$ into $\mathcal{B}(\hat{\mathcal{H}}(\infty))$ is locally Hölder continuous with exponent γ that satisfies $\gamma < 1$, $\gamma < (s - 1/2)$, if $\lambda \neq \pm\ c_+\ \rho_j^E$, $\lambda \neq \pm\ c_+\ \rho_j^M$, for $j = 1, 2, 3, \cdots$, and $\gamma < 1/2$, $\gamma < (s - 1/2)$, if $\lambda = \pm\ c_+\ \rho_j^E$, or $\lambda = \pm\ c_+\ \rho_j^M$, for some $j = 1, 2, 3, \cdots$, where $s = \frac{1+\delta}{2}$.

Proof: This lemma is proven as Lemma 7.4 in Chapter 2.

Q. E. D.

We proceed to study the continuity of the wave and scattering operators in the perturbation.

Remark 4.5

We consider for $i = 1, 2, 3, \cdots$, real valued functions on $\mathbf{R}^3$, $\epsilon_i(x,z)$, $\mu_i(x,z)$ that satisfy (3.4), (3.24), and (3.25), and such that

$$\lim_{i\to\infty} \left\| (1+|x|+|z|)^{1+\delta} \, |\epsilon_i - \epsilon| \, \right\|_\infty = 0, \tag{4.23}$$

$$\lim_{i\to\infty} \left\| (1+|x|+|z|)^{1+\delta} \, |\mu_i - \mu| \, \right\|_\infty = 0. \tag{4.24}$$

Let us denote by $W_i^{\pm}$, and S_i the wave and scattering operators for H_{ϵ_i,μ_i} defined as in (3.6) with $\epsilon_i(x,z)$, and $\mu_i(x,z)$ instead of $\epsilon(x,z)$ and $\mu(x,z)$.

We also define

$$\tilde{W}^{\pm} = L\, W^{\pm}, \quad \tilde{W}_i^{\pm} = L_i\, W_i^{\pm}, \tag{4.25}$$

where L_i is defined as in (3.11), (3.12) with $\epsilon_i(x,z)$, $\mu_i(x,z)$, $i = 1, 2, 3, \cdots$, instead of $\epsilon(x,z)$, $\mu(x,z)$. Then in the strong topology in $\mathcal{H}_0$

$$s - \lim_{i\to\infty} \tilde{W}_i^{\pm} = \tilde{W}^{\pm}, \tag{4.26}$$

and

$$s - \lim_{i\to\infty} S_i = S. \tag{4.27}$$

Proof: This is proven as Remark 7.5 in Chapter 2.

Remark 4.6

Let us denote by $S_{\epsilon,\mu}(\lambda)$ the operator (4.16) corresponding to $H_{\epsilon,\mu}$.

Then for any compact set $K \subset \mathbf{R} \setminus \sigma_p(H_{\epsilon,\mu})$ there is a neighborhood, $O_{\epsilon,\mu}$, of ϵ, μ such that for $\tilde{\epsilon}, \tilde{\mu} \in O_{\epsilon,\mu}$, $K \subset \mathbf{R} \setminus \sigma_p(H_{\tilde{\epsilon},\tilde{\mu}})$ and

$$\left\| S_{\epsilon,\mu}(\lambda) - S_{\tilde{\epsilon},\tilde{\mu}}(\lambda) \right\|_{\mathcal{B}(\hat{\mathcal{H}}(\lambda))} \leq C \left\| (1+|x|+|z|)^{1+\delta} (|\epsilon-\tilde{\epsilon}| + |\mu-\tilde{\mu}|) \right\|_\infty, \tag{4.28}$$

for some constant C, and all $\lambda \in K$.

Proof: This remark follows as Remark 7.6 of Chapter 2.

Remark 4.7

For any $v(\lambda) \in \hat{\mathcal{H}}(\lambda)$, where

$$v(\lambda) = v_0(\omega) \bigoplus_{j=1}^{\ell_E} v_j^E(\nu) \bigoplus_{j=1}^{\ell_M} v_j^M(\nu), \tag{4.29}$$

for some nonnegative integers $\ell_E,\ \ell_M$, we define

$$\phi_{0,v_0}(x,z,\lambda) = \int_{S_c} \phi_0(x,z,\lambda,\omega)\ v_0(\omega)d\omega, \tag{4.30}$$

$$\phi_{j,v_j^E}^E(x,z,\lambda) = \int_{S_1^1} \phi_j^E(x,z,\lambda,\nu)\ v_j^E(\nu)d\nu, \tag{4.31}$$

$$\phi_{j,v_j^M}^M(x,z,\lambda) = \int_{S_1^1} \phi_j^M(x,z,\lambda,\nu)\ v_j^M(\nu)d\nu. \tag{4.32}$$

We remark that

$$B^*(\lambda)v = \eta_{-s}\Big(\phi_{0,v_0} + \sum_{j=1}^{\ell_E} \phi_{j,v_j^E}^E + \sum_{j=1}^{\ell_M} \phi_{j,v_j^M}^M\Big). \tag{4.33}$$

Let us denote by

$$\phi_{0,v_0}^-(x,z,\lambda) = \phi_{0,v_0} + \lambda\ R_+(\lambda)\ (I-e^2)\ \phi_{0,v_0}, \tag{4.34}$$

$$\phi_{j,v_j^E}^{-,E}(x,z,\lambda) = \phi_{j,v_j^E}^E + \lambda\ R_+(\lambda)\ (I-e^2)\ \phi_{j,v_j^E}^E, \tag{4.35}$$

$$\phi_{j,v_j^M}^{-,M}(x,z,\lambda) = \phi_{j,v_j}^M + \lambda\ R_+(\lambda)\ (I-e^2)\ \phi_{j,v_j}^M. \tag{4.36}$$

Let us denote

$$T(\lambda) = S(\lambda) - I. \tag{4.37}$$

Then for all $u,\ v \in \hat{\mathcal{H}}(\lambda)$

$$(T(\lambda)u,\ v)_{\hat{\mathcal{H}}(\lambda)} = -2\pi i\lambda\Big((I-e^{-2})\Big(\phi_{0,u_0}^- + \sum_{j=1}^{\ell_E} \phi_{j,u_j^E}^{-,E} +$$

$$+ \sum_{j=1}^{\ell_M} \phi_{j,u_j^M}^{-,M}\Big),\Big(\phi_{0,v_0} + \sum_{j=1}^{\ell_E} \phi_{j,v_j^M}^E + \cdot \sum_{j=1}^{\ell_M} \phi_{j,v_j^M}^M\Big)\Big)_{\mathcal{H}_0}. \tag{4.38}$$

If the following decay conditions are valid

$$|\epsilon(x,z)-\epsilon_0(z)|\le C\,(1+|x|+|z|)^{-3-\delta}, \tag{4.39}$$

$$|\mu(x,z)-\mu_0(z)|\le C\,(1+|x|+|z|)^{-3-\delta}, \tag{4.40}$$

for some **C**, $\delta>0$, the order of integration in (4.38) can be inverted and $T(\lambda)$ becomes an integral operator of Hilbert-Schmid type:

$$(T(\lambda)u)_0(\omega) = -2\pi i\lambda \left[\int_{S_c} T_{0,0}(\lambda,\omega,\acute{\omega})u_0(\acute{\omega})d\,\acute{\omega} + \right.$$

$$\left. + \sum_k \int_{S_1^1} T_{0,k}^E(\lambda,\omega,\nu)u_k^E(\nu)d\nu + \sum_k \int_{S_1^1} T_{0,k}^M(\lambda,\omega,\nu)u_k^M(\nu)d\nu \right], \tag{4.41}$$

$$(T(\lambda)u)_j^E(\nu) = -2\pi i\lambda \left[\int_{S_c} T_{j,0}^E(\lambda,\nu,\omega)u_0(\omega)d\omega + \right.$$

$$\left. + \sum_k \int_{S_1^1} T_{j,k}^{E,E}(\lambda,\nu,\acute{\nu})u_k^E(\acute{\nu})d\,\acute{\nu} + \sum_k \int_{S_1^1} T_{j,k}^{E,M}(\lambda,\nu,\acute{\nu})u_k^M(\acute{\nu})d\,\acute{\nu} \right], \tag{4.42}$$

$$(T(\lambda)u)_j^M(\nu) = -2\pi i\lambda \left[\int_{S_c} T_{j,0}^M(\lambda,\nu,\omega)u_0(\omega)d\omega + \right.$$

$$\left. + \sum_k \int_{S_1^1} T_{j,k}^{M,M}\;(\lambda,\nu,\acute{\nu})v_k^M(\acute{\nu})d\,\acute{\nu} + \sum_k \int_{S_1^1} T_{j,k}^{M,E}(\lambda,\nu,\acute{\nu})v_k^E(\acute{\nu})d\,\acute{\nu} \right]. \tag{4.43}$$

Where

$$T_{0,0}(\lambda,\omega,\acute{\omega}) = \Big((I-e^{-2})\phi_0^-(x,z,\lambda,\acute{\omega}),\ \phi_0(x,z,\lambda,\omega)\Big)_{\mathcal{H}_0}, \tag{4.44}$$

$$T_{0,k}^E(\lambda,\omega,\nu) = \Big((I-e^{-2})\phi_k^{-,E}(x,z,\lambda,\nu),\ \phi_0(x,z,\lambda,\omega)\Big)_{\mathcal{H}_0}, \tag{4.45}$$

$$T_{0,k}^M(\lambda,\omega,\nu) = \Big((I-e^{-2})\phi_k^{-,M}(x,z,\lambda,\nu),\ \phi_0(x,z,\lambda,\omega)\Big)_{\mathcal{H}_0}, \tag{4.46}$$

$$T_{k,0}^E(\lambda,\nu,\omega) = \Big((I-e^{-2})\phi_0^-(x,z,\lambda,\omega),\ \phi_k^E(x,z,\lambda,\nu)\Big)_{\mathcal{H}_0}, \tag{4.47}$$

$$T_{k,j}^{E,E}(\lambda,\nu,\acute{\nu}) = \Big((I-e^{-2})\phi_j^{-,E}(x,z,\lambda,\acute{\nu}),\ \phi_k^E(x,z,\lambda,\nu)\Big)_{\mathcal{H}_0}, \tag{4.48}$$

$$T_{k,j}^{E,M}(\lambda,\nu,\acute{\nu}) = \Big((I-e^{-2})\phi_j^{-,M}(x,z,\lambda,\acute{\nu}),\ \phi_k^E(x,z,\lambda,\nu)\Big)_{\mathcal{H}_0}, \tag{4.49}$$

$$T_{k,0}^{M}(\lambda,\nu,\omega) = \Big((I-e^{-2})\phi_0^-(x,z,\lambda,\omega),\ \phi_k^M(x,z,\lambda,\nu)\Big)_{\mathcal{H}_0}, \quad (4.50)$$

$$T_{k,j}^{M,M}(\lambda,\nu,\acute{\nu}) = \Big((I-e^{-2})\phi_j^{-,M}(x,z,\lambda,\acute{\nu}),\ \phi_k^M(x,z,\lambda,\nu)\Big)_{\mathcal{H}_0}, \quad (4.51)$$

$$T_{k,j}^{M,E}(\lambda,\nu,\acute{\nu}) = \Big((I-e^{-2})\phi_j^{-,E}(x,z,\lambda,\acute{\nu}),\ \phi_k^M(x,z,\lambda,\nu)\Big)_{\mathcal{H}_0}. \quad (4.52)$$

Where

$$\phi_0^-(x,z,\lambda,\omega) = \phi_0(x,z,\lambda,\omega) + \lambda\ R_+(\lambda)(I-e^2)\phi_0, \quad (4.53)$$

$$\phi_k^{-,E}(x,z,\lambda,\nu) = \phi_k^E(x,z,\lambda,\nu) + \lambda\ R_+(\lambda)(I-e^2)\phi_k^E, \quad (4.54)$$

$$\phi_k^{-,M}(x,z,\lambda,\nu) = \phi_k^M(x,z,\lambda,\nu) + \lambda\ R_+(\lambda)(I-e^2)\phi_k^M, \quad (4.55)$$

$k = 1, 2, 3, \cdots$.

Note that

$$H\ \phi_0^-(x,z,\lambda,\omega) = \lambda\ \phi_0^-(x,z,\lambda,\omega), \quad (4.56)$$

$$H\ \phi_k^{-,E}(x,z,\lambda,\nu) = \lambda\ \phi_k^{-,E}(x,z,\lambda,\nu), \quad (4.57)$$

and

$$H\ \phi_k^{-,M}(x,z,\lambda,\nu) = \lambda\ \phi_k^{-,M}(x,z,\lambda,\nu). \quad (4.58)$$

Appendix 1

In this appendix we construct the eigenfunctions and generalized eigenfunctions of the reduced acoustic propagator, and we prove the existence and Hölder continuity of the trace maps.

We denote

$$c_0(y) = \begin{cases} c_+, & y \geq h, \\ c_h, & 0 \leq y < h, \\ c_-, & y < 0, \end{cases} \tag{1.1}$$

$$\mu_0(y) = \begin{cases} \mu_+, & y \geq h, \\ \mu_h, & 0 \leq y < h, \\ \mu_-, & y < 0, \end{cases} \tag{1.2}$$

where c_+, c_h, c_-, μ_+, μ_h, μ_-, and h are positive constants, $c_+ \leq c_-$, and $c_h < c_+$.

Let $\mathcal{L}$ be the Hilbert space consisting of all complex valued Lebesgue square integrable functions on $\mathbf{R}$ with scalar product given by

$$(\varphi, \psi)_{\mathcal{L}} = \int \varphi(y)\overline{\psi}(y)\, c_0^{-2}(y)\mu_0^{-1}(y)dy. \tag{1.3}$$

For $\rho > 0$ the reduced acoustic propagator is the following operator in $\mathcal{L}$,

$$A(\rho)\varphi = c_0^2(y)\Big[-\mu_0(y)\,\frac{d}{dy}\left(\frac{1}{\mu_0(y)}\,\frac{d}{dy}\,\varphi\right) + \rho^2\varphi\Big], \tag{1.4}$$

with domain

$$D(A(\rho)) = H_1(\mathbf{R}) \cap \Big\{\varphi\, :\, \frac{d}{dy}\left(\frac{1}{\mu_0(y)}\,\frac{d\varphi}{dy}\right) \in \mathcal{L}\Big\}. \tag{1.5}$$

Then $A(\rho)$ is selfadjoint and $A(\rho) \geq c_h^2\rho^2$ (see Wilcox 1984, Chapter 3, Section 2).

Moreover, (see Wilcox 1984, Chapter 3, Section 4),

$$\sigma_c(A(\rho)) = \sigma_e(A(\rho)) = [c_+^2\rho^2, \infty), \tag{1.6}$$

$$\sigma_p(A(\rho)) \subset [c_h^2\rho^2,\ c_+^2\rho^2]. \tag{1.7}$$

If λ is an eigenvalue with eigenvector φ, then

$$A(\rho)\varphi = \lambda\varphi, \tag{1.8}$$

and since $\varphi \in D(A(\rho))$ we must have that

$$\varphi(h_+) = \varphi(h_-), \quad \mu_+^{-1}\ \varphi'(h_+) = \mu_h^{-1}\ \varphi'(h_-),$$

$$\varphi(0_+) = \varphi(0_-), \quad \mu_h^{-1}\ \varphi'(0_+) = \mu_-^{-1}\ \varphi'(0_-),$$

where $\varphi'(y) = \frac{d}{dy}\varphi(y)$, and for $a \in \mathbf{R}$, $\varphi(a_+)$, $\varphi(a_-)$, $\varphi'(a_+)$, *and* $\varphi'(a_-)$ are respectively the limits from the right and the left at $y = a$, of $\varphi(y)$ and $\varphi'(y)$. Note that it follows by elliptic regularity that these limits exist for $\varphi(y)$.

It is easily checked that at $\lambda = c_+^2\ \rho^2$, and $\lambda = c_h^2\ \rho^2$ the eigenvalue equation (1.8) has no solution. Then $\sigma_p(A(\rho)) \subset (c_h^2\ \rho^2,\ c_+^2\ \rho^2)$, and by Lemma 4.3 in chapter 3 of Wilcox 1984 the eigenvalues of $A(\rho)$ are all simple.

We compute the eigenvalues and eigenvectors of $A(\rho)$ as in Wilcox 1976. However, since that paper considers the case of the half line $(0, \infty)$ with Dirichlet boundary condition at $y = 0$, and we study the case of the full line $\mathbf{R}$, we give below the main steps in the solution of the eigenvalues and eigenvectors problem.

For $c_h^2\ \rho^2 < \lambda < c_+^2\ \rho^2$, we can write the solution in $\mathcal{L}$ of (1.8) as

$$\varphi(y) = \begin{cases} e^{-\tilde{\lambda}_+(y-h)} & , \quad y \geq h, \\ A\cos(\lambda_h(y-h)) + B\sin(\lambda_h(y-h)) & , \quad 0 \leq y \leq h, \\ C\ e^{\tilde{\lambda}_- y} & , \quad y \leq 0, \end{cases} \tag{1.9}$$

where

$$\tilde{\lambda}_+ = (\rho^2 - \lambda\ c_+^{-2})^{1/2},\ \tilde{\lambda}_- = (\rho^2 - \lambda\ c_-^{-2})^{1/2}, \tag{1.10}$$

$$\lambda_h = (\lambda\ c_h^{-2} - \rho^2)^{1/2}, \tag{1.11}$$

where the square roots are positive. Then $\varphi(h_+) = \varphi(h_-)$ implies $A = 1$, and $\mu_+^{-1} \varphi'(h_+) = \mu_h^{-1} \varphi'(h_-)$, implies $B = -\mu_h \tilde{\lambda}_+ (\lambda_h \mu_+)^{-1}$.

The conditions $\varphi(0_-) = \varphi(0_+)$, and $\mu_-^{-1} \varphi'(0_-) = \mu_h^{-1} \varphi'(0_+)$ imply that

$$tang(\lambda_h h) = \frac{\left(\frac{\mu_h}{\mu_+}\frac{\tilde{\lambda}_+}{\lambda_h} + \frac{\mu_h}{\mu_-}\frac{\tilde{\lambda}_-}{\lambda_h}\right)}{\left(1 - \frac{\mu_h}{\mu_+}\frac{\tilde{\lambda}_+}{\lambda_h} \cdot \frac{\mu_h}{\mu_-}\frac{\tilde{\lambda}_-}{\lambda_h}\right)}, \tag{1.12}$$

and then

$$\lambda_h h = arctang\Big(\frac{\mu_h}{\mu_+}\frac{\tilde{\lambda}_+}{\lambda_h}\Big) + arctang\Big(\frac{\mu_h}{\mu_-}\frac{\tilde{\lambda}_-}{\lambda_h}\Big) + (j-1)\pi, \tag{1.13}$$

where $j = 1, 2, 3, \cdots$, and where we take the principal branch of the function *arctang*.

We define the function

$$G(\lambda, \rho) = \frac{1}{\pi}\left[\lambda_h h - arctang\Big(\frac{\mu_h}{\mu_+}\frac{\tilde{\lambda}_+}{\lambda_h}\Big) - arctang\Big(\frac{\mu_h}{\mu_-}\frac{\tilde{\lambda}_-}{\lambda_h}\Big)\right]. \tag{1.14}$$

Then, as it is easily checked, for $c_h^2 \rho^2 < \lambda < c_+^2 \rho^2$,

$$\frac{\partial G}{\partial \lambda} > 0, \tag{1.15}$$

and moreover,

$$G(c_h^2 \rho^2,\ \rho) = -1, \tag{1.16}$$

$$G(c_+^2 \rho^2,\ \rho) = \frac{1}{\pi}\left[h\,\rho\Big(\frac{c_+^2}{c_h^2} - 1\Big)^{1/2} - arctang\left(\frac{\mu_h}{\mu_-}\frac{\Big(1 - \frac{c_+^2}{c_-^2}\Big)^{1/2}}{\Big(\frac{c_+^2}{c_h^2} - 1\Big)^{1/2}}\right)\right]. \tag{1.17}$$

Then for each fixed ρ and j there can be at most one solution of (1.13) for λ in the interval $(c_h^2 \rho^2,\ c_+^2 \rho^2)$. If it exists we denote this solution by $\lambda_j(\rho)$. By (1.17) the solution exists if and only if $\rho > \rho_j$, where for $j = 1, 2, 3, \cdots$,

$$\rho_j = \frac{1}{h\Big(\frac{c_+^2}{c_h^2} - 1\Big)^{1/2}}\left[arctang\left(\frac{\mu_h}{\mu_-}\frac{\Big(1 - \frac{c_+^2}{c_-^2}\Big)^{1/2}}{\Big(\frac{c_+^2}{c_h^2} - 1\Big)^{1/2}}\right) + (j-1)\pi\right]. \tag{1.18}$$

Then the functions $\lambda_j(\rho)$ are defined for $\rho \in U_j = (\rho_j, \infty)$, and by the implicit function theorem are analytic functions of ρ. Note that $\rho_1 \geq 0$, with $\rho_1 = 0$ if and only if $c_+ = c_-$, and that $\rho_{j+1} > \rho_j$, $\lim_{j \to \infty} \rho_j = \infty$. Moreover if $j > i$, for $\rho > \rho_j$

$$\lambda_j(\rho) > \lambda_i(\rho). \tag{1.19}$$

Furthermore for $\rho \in U_j$

$$G(\lambda_j(\rho),\ \rho) = j - 1, \tag{1.20}$$

then

$$\frac{\partial G}{\partial \lambda}(\lambda_j(\rho),\ \rho)\ \frac{d\lambda_j(\rho)}{d\rho} + \frac{\partial G}{\partial \rho}\ (\lambda_j(\rho), \rho) = 0, \tag{1.21}$$

and since by explicit calculation

$$\frac{\partial G}{\partial \rho}(\lambda_j(\rho),\ \rho) < 0, \tag{1.22}$$

it follows from (1.15) that

$$\frac{d\lambda_j(\rho)}{d\rho} > 0, \quad \text{for} \quad \rho \in U_j. \tag{1.23}$$

As $\lambda_j(\rho) > c_h^2 \rho^2$, it follows from (1.18) and (1.20) that

$$\lim_{\rho \downarrow \rho_j} \lambda_j(\rho) = c_+^2 \rho_j^2. \tag{1.24}$$

Let us construct a parametric representation for the $\lambda_j(\rho)$. Denote

$$\beta = \frac{\rho^2}{\lambda_j(\rho)}, \tag{1.25}$$

for $\rho \in U_j$. Then

$$\frac{1}{c_+^2} < \beta < \frac{1}{c_h^2}, \tag{1.26}$$

and by (1.13) and (1.25)

$$\rho(\beta) = \frac{\beta^{1/2}}{h(c_h^{-2} - \beta)^{1/2}} \left[arctang\left(\frac{\mu_h}{\mu_+} \frac{(\beta - c_+^{-2})^{1/2}}{(c_h^{-2} - \beta)^{1/2}} \right) + \right.$$

$$\left. + arctang\left(\frac{\mu_h}{\mu_+} \frac{(\beta - c_-^{-2})^{1/2}}{(c_h^{-2} - \beta)^{1/2}} \right) + (j-1)\pi \right], \tag{1.27}$$

$$\omega_j(\rho) \ = \ \sqrt{\lambda_j(\rho)} \ = \ \rho(\beta)\beta^{-1/2}. \tag{1.28}$$

By (1.27)

$$\frac{d\rho(\beta)}{d\beta} > 0. \tag{1.29}$$

It follows that the inverse function $\beta = \beta(\rho)$ exists for $\rho \in (\rho_j, \infty)$ and it is strictly increasing.

Moreover by (1.27)

$$\lim_{\beta \uparrow c_h^{-2}} \rho(\beta) = \infty. \tag{1.30}$$

Since by (1.25)

$$\frac{1}{\rho}\, \omega_j(\rho) = \frac{1}{\beta^{1/2}}, \tag{1.31}$$

$\frac{1}{\rho}\, \omega_j(\rho)$ is a decreasing function of ρ, and

$$c_h < \frac{1}{\rho}\, \omega_j(\rho) < c_+, \quad \rho \in U_j. \tag{1.32}$$

Furthermore,

$$\lim_{\rho \to \infty} \frac{1}{\rho}\, \omega_j(\rho) = c_h. \tag{1.33}$$

Let us extend $\omega_j(\rho)$ to $\rho = \rho_j$ by defining (see (1.24))

$$\omega_j(\rho_j) = c_+ \rho_j. \tag{1.34}$$

We already know that $\omega_j(\rho)$ is an analytic function for $\rho \in U_j$. Moreover it follows from (1.27) and (1.28) that $\omega_j(\rho)$ is also analytic at $\rho = \rho_j$. Furthermore it follows from (1.21) and explicit calculation that

$$\lim_{\rho \downarrow \rho_j} \frac{d\omega_j(\rho)}{d\rho} = - \lim_{\rho \downarrow \rho_j} \frac{\frac{\partial G}{\partial \rho}(\omega_j^2(\rho),\, \rho)}{\frac{\partial G}{\partial \lambda}(\omega_j^2(\rho),\, \rho) 2\, \omega_j(\rho)} = c_+. \tag{1.35}$$

Let us summarize the properties of the $\omega_j(\rho)$ that we have proved.

1. For $j = 1, 2, 3, \cdots$, the functions $\omega_j(\rho)$ are analytic functions of $\rho \in [\rho_j,\, \infty)$ into $\mathbf{R}^+$, where $\rho_1 \geq 0$, $\rho_{j+1} > \rho_j$, and $\lim_{j \to \infty} \rho_j = \infty$. Moreover $\rho_1 = 0$ if and only if $c_- = c_+$.

2. For $\rho \in [\rho_{j+1}, \infty)$, $\omega_{j+1}(\rho) > \omega_j(\rho)$. (1.36)

3. For $\rho \in (\rho_j, \infty)$, $c_h \rho < \omega_j(\rho) < c_+ \rho$. (1.37)

4. $\lim_{\rho \downarrow \rho_j} \omega_j(\rho) = c_+ \rho_j$, $\lim_{\rho \to \infty} \frac{1}{\rho}\, \omega_j(\rho) = c_h$, (1.38)

and $\frac{\omega_j(\rho)}{\rho}$ is a decreasing function of ρ.

5. For $\rho > \rho_j$, $\frac{d\omega_j}{d\rho} > 0$, and $\left.\frac{d\omega_j}{d\rho}\right)_{\rho = \rho_j} = c_+$. (1.39)

6. For $\rho_j < \rho \leq \rho_{j+1}$, $A(\rho)$ has exactly j eigenvalues of multiplicity one given by $\omega_1^2(\rho) < \omega_2^2(\rho) < \cdots < \omega_j^2(\rho)$.

Let us denote by $\psi_j(y, \rho)$ the normalized eigenvector corresponding to $\omega_j^2(\rho)$. Then by (1.9) and the conditions at $y = 0$, and $y = h$ stated below (1.8) we have that

$$\psi_j(y, \rho) = a_j(\rho) \begin{cases} e^{-\tilde{\lambda}_{+,j}(y-h)}, & y \geq h, \\ \cos\left(\lambda_{h,j}(y-h)\right) - \frac{\mu_h\, \tilde{\lambda}_{+,j}}{\mu_+\, \lambda_{h,j}} \sin\left(\lambda_{h,j}(y-h)\right), & 0 \leq y \leq h, \\ e^{\tilde{\lambda}_{-,j} y}\left(\cos(\lambda_{h,j} h) + \frac{\mu_h\, \tilde{\lambda}_{+,j}}{\lambda_h\, \mu_+} \sin(\lambda_{h,j})\right), & y \leq 0, \end{cases} \tag{1.40}$$

where $\tilde{\lambda}_{+,j} = (\rho^2 - \omega_j^2(\rho)c_+^{-2})^{1/2}$, $\tilde{\lambda}_{-,j} = (\rho^2 - \omega_j^2(\rho)c_-^{-2})^{1/2}$, $\lambda_{h,j} = (\omega_j^2(\rho)c_h^{-2} - \rho^2)^{1/2}$, and where the constant $a_j(\rho)$ is chosen such that

$$\left\| \psi_j(y,\rho) \right\|_{\mathcal{L}} = 1. \tag{1.41}$$

It follows by explicit computation using (1.40) that

$$a_j(\rho) = (\tilde{\lambda}_{+,j}(\rho))^{1/2} J(\rho), \tag{1.42}$$

where $J(\rho)$ is continuous and strictly positive on $[\rho_j, \infty)$.

We now construct the generalized eigenfunctions of $A(\rho)$ following Wilcox 1984.

For $\lambda > c_-^2 \rho^2$ we denote

$$\lambda_{\pm} = (\lambda\, c_{\pm}^{-2} - \rho^2)^{1/2}. \tag{1.43}$$

We define

$$\begin{aligned} \phi_1(y,\rho,\lambda) = e^{i\lambda_+ h} \Bigg[& \left(\cos(\lambda_h h) - i\,\frac{\mu_h\,\lambda_+}{\mu_+\,\lambda_h}\,\sin(\lambda_0 h)\right)\cos(\lambda_- y) + \\ & + \left(i\,\frac{\mu_-\,\lambda_+}{\mu_+\,\lambda_-}\,\cos(\lambda_h h) + \right. \\ & \left. + \frac{\mu_-\,\lambda_h}{\mu_h\,\lambda_-}\,\sin(\lambda_h h)\right)\sin(\lambda_- y)\Bigg], \qquad y \le 0, \end{aligned} \tag{1.44}$$

$$\begin{aligned} \phi_1(y,\rho,\lambda) &= e^{i\lambda_+ h}\Big(\cos\left(\lambda_h(y-h)\right) + \\ &\quad + i\,\frac{\mu_h\,\lambda_+}{\mu_+\,\lambda_h}\,\sin\left(\lambda_h(y-h)\right)\Big),\ 0 \le y \le h, \end{aligned} \tag{1.45}$$

$$\phi_1(y,\rho,\lambda) = e^{i\lambda_+ y}, \qquad y \ge h, \tag{1.46}$$

and

$$\phi_4(y,\rho,\lambda) = e^{-i\lambda_- y}, \qquad y \le 0, \tag{1.47}$$

$$\phi_4(y,\rho,\lambda) = \cos(\lambda_h y) - i\,\frac{\mu_h\,\lambda_-}{\mu_-\,\lambda_h}\,\sin(\lambda_h y),\ 0 \le y \le h, \tag{1.48}$$

$$\begin{aligned} \phi_4(y,\rho,\lambda) &= \cos\left(\lambda_+(y-h)\right)\left[\cos(\lambda_h h) - i\,\frac{\mu_h\,\lambda_-}{\mu_-\,\lambda_h}\,\sin(\lambda_0 h)\right] - \\ &\quad - \frac{\mu_+}{\lambda_+}\,\sin\left(\lambda_+(y-h)\right)\left[\frac{\lambda_h}{\mu_h}\,\sin(\lambda_h h) + i\,\frac{\lambda_-}{\mu_-}\,\cos(\lambda_h h)\right],\ y \ge h, \end{aligned} \tag{1.49}$$

where λ_h is as in (1.11).

It is easily checked that $\phi_1(y, \rho, \lambda)$ and $\phi_4(y, \rho, \lambda)$ are solutions in distribution sense of the equation (1.8) (note that they satisfy the conditions at $y = 0$ and $y = h$ stated below (1.8)).

Moreover by Corollary 3.5, Chapter 3 of Wilcox 1984 they coincide with the functions $\phi_1(y, \rho, \lambda)$ and $\phi_4(y, \rho, \lambda)$ of Theorem 3.1 Chapter 3 of Wilcox 1984.

The generalized eigenfunctions $\psi_\pm(y, \rho, \lambda)$ are defined in Wilcox 1984, page 24, formula (1.16) as

$$\psi_+(y, \rho, \lambda) = a_+(\rho, \lambda)\ \phi_4(y, \rho, \lambda), \tag{1.50}$$

$$\psi_-(y, \rho, \lambda) = a_-(\rho, \lambda)\ \phi_1(y, \rho, \lambda), \tag{1.51}$$

for $\lambda > c_-^2 \rho^2$, where the normalization coefficients are defined as

$$a_\pm(\rho, \lambda) = \left[\frac{\lambda_\pm}{\pi\ \mu_\pm |[\phi_1, \phi_4]|^2} \right]^{1/2}, \tag{1.52}$$

where

$$[\phi_1, \phi_4] = \phi_1\ \mu^{-1}\ \phi_4' - \phi_1'\ \mu^{-1}\ \phi_4, \tag{1.53}$$

is the Wronskian of ϕ_1 and ϕ_4. Note that since ϕ_1 and ϕ_4 are linearly independent $[\phi_1, \phi_4] \neq 0$, and that (1.53) is independent of y. Formula (1.52) follows from equations (6.10) and (6.14) on pages 72 and 73 of Wilcox 1984. Note that we have taken $\Theta_\pm = 0$ in equation (6.14).

Since ϕ_1 and ϕ_4 are explicitly known we can evaluate (1.52). The result is

$$\begin{aligned} a_\pm(\rho, \lambda) = & \left(\frac{\lambda_\pm}{\pi\ \mu_\pm} \right)^{\frac{1}{2}} \left[\left(\frac{\lambda_-}{\mu_-} + \frac{\lambda_+}{\mu_+} \right)^2 (\cos(\lambda_h h))^2 + \right. \\ & \left. + \left(\frac{\lambda_h}{\mu_h} + \frac{\mu_h\ \lambda_+\ \lambda_-}{\mu_+\ \mu_-\ \lambda_h} \right)^2 (\sin(\lambda_h h))^2 \right]^{-\frac{1}{2}}. \end{aligned} \tag{1.54}$$

For $c_+^2\ \rho^2 < \lambda < c_-^2\ \rho^2$ denote

$$\tilde{\lambda}_- = (\rho^2 - \lambda\ c_-^{-2})^{1/2}, \tag{1.55}$$

and

$$\phi_3(y, \rho, \lambda) \ = \ e^{\tilde{\lambda}_- y} \quad , \quad y \leq 0, \tag{1.56}$$

$$\phi_3(y,\rho,\lambda) = \cos(\lambda_h(y-h))\Big(\cos(\lambda_h h) + \frac{\mu_h\,\tilde{\lambda}_-}{\mu_-\,\lambda_h}\sin(\lambda_h h)\Big) +$$

$$+ \sin(\lambda_h(y-h))\Big(-\sin(\lambda_h h) + \frac{\mu_h\,\tilde{\lambda}_-}{\mu_-\,\lambda_h}\cos(\lambda_h h)\Big), \; 0 \le y \le h, \tag{1.57}$$

$$\phi_3(y,\rho,\lambda) = \cos(\lambda_+(y-h))\Big(\cos(\lambda_h h) + \frac{\mu_h\,\tilde{\lambda}_-}{\mu_-\,\lambda_h}\sin(\lambda_h h)\Big) +$$

$$+ \frac{\mu_+}{\lambda_+}\sin(\lambda_+(y-h))\Big(\frac{\tilde{\lambda}_-}{\mu_-}\cos(\lambda_h h) - \frac{\lambda_h}{\mu_h}\sin(\lambda_h h)\Big), \quad y \ge h. \tag{1.58}$$

It is easily checked that $\phi_3(y,\rho,\lambda)$ is a solution in distribution sense of equation (1.8). Moreover by Corollary 3.5, Chapter 3 of Wilcox 1984, it coincides with the function $\phi_3(y,\rho,\lambda)$ of Theorem 3.1, Chapter 3 of Wilcox 1984.

If $c_+ < c_-$ the generalized eigenfunction $\psi_0(y,\rho,\lambda)$ is defined as (see Wilcox 1984, page 24, formula (1.15))

$$\psi_0(y,\rho,\lambda) = a_0(\rho,\lambda)\;\phi_3(y,\rho,\lambda), \tag{1.59}$$

where (see Wilcox 1984, page 74, formulas (6.19) and (6.21))

$$a_0(\rho,\lambda) = \left[\frac{\lambda_+}{\pi\;\mu_+|[\phi_1,\phi_3]|^2}\right]^{1/2}. \tag{1.60}$$

Note that we have taken $\Theta_0(\rho,\lambda) = 0$ in (6.21), page 74 of Wilcox 1984. By explicit computation we obtain that

$$a_0(\rho,\lambda) = \left(\frac{\lambda_+}{\pi\;\mu_+}\right)^{1/2}\left[\frac{\lambda_+^2}{\mu_+^2}\left(\cos(\lambda_h h) + \frac{\mu_h\,\tilde{\lambda}_-}{\mu_-\,\lambda_h}\sin(\lambda_h h)\right)^2 + \right.$$

$$\left. + \left(\frac{\tilde{\lambda}_-}{\mu_-}\cos(\lambda_h h) - \frac{\lambda_h}{\mu_h}\sin(\lambda_h h)\right)^2\right]^{-\frac{1}{2}}. \tag{1.61}$$

Let the sets Ω, Ω_j, $j = 0,1,2,\cdots$, and S_c, S_+, S_0, S_-, be defined respectively as in formulas (1.30) to (1.36) of Section 1, Chapter 2. Also let the analytic bijections χ_+, χ_0, and χ_- be defined as in formulas (1.37) to (1.43) of Section 1, Chapter 2.

We define the following composite generalized eigenfunction

$$\phi_0(x,y,\lambda,\omega) = c_+^{1/2}\;\lambda^{n/4}\;\gamma_+^{1/2}\;\frac{1}{(2\pi)^{n/2}}\,e^{ik\cdot x}\;\psi_+(y,|k|,\lambda), \tag{1.62}$$

where $(\lambda, k) = \chi_+^{-1}(\lambda, \omega)$, for $(\lambda, \omega) \in (0, \infty) \times S_+$,

$$\phi_0(x, y, \lambda, \omega) = c_+^{1/2} \ \lambda^{n/4} \ \gamma_+^{1/2} \ \frac{1}{(2\pi)^{n/2}} \ e^{ik \cdot x} \ \psi_0(y, |k|, \lambda), \tag{1.63}$$

where $(\lambda, k) = \chi_0^{-1}(\lambda, \omega)$, for $(\lambda, \omega) \in (0, \infty) \times S_0$,

$$\phi_0(x, y, \lambda, \omega) = c_-^{1/2} \ \lambda^{n/4} \ \gamma_-^{1/2} \ \frac{1}{(2\pi)^{n/2}} \ e^{ik \cdot x} \ \psi_-(y, |k|, \lambda), \tag{1.64}$$

where $(\lambda, k) = \chi_-^{-1}(\lambda, \omega)$, for $(\lambda, \omega) \in (0, \infty) \times S_-$.

Let us denote by $O_j = (\lambda_j, \infty)$, with $\lambda_j = c_+^2 \ \rho_j^2, \quad j = 1, 2, 3, \cdots$. Let $\beta_j(\lambda)$ be the inverse function to $\lambda_j(\rho)$ (recall that $\frac{d}{d\rho} \ \lambda_j(\rho) > 0$). That is to say $\beta_j(\lambda)$ is a function from $\overline{O}_j$ onto $\overline{U}_j$ such that

$$\beta_j(\lambda) = \rho \iff \lambda = \lambda_j(\rho), \tag{1.65}$$

for $\lambda \in \overline{O}_j, \quad \rho \in \overline{U}_j$.

We define for $\lambda \in O_j, \quad \nu \in S_1^{n-1}$

$$\phi_j(x, y, \lambda, \nu) \ = \ (\beta_j(\lambda))^{\frac{n-1}{2}} \ (\beta_j'(\lambda))^{1/2} \ \frac{1}{(2\pi)^{n/2}} \ e^{i\beta_j(\lambda)\nu \cdot x} \ \psi_j(y, \beta_j(\lambda)), \tag{1.66}$$

for $j = 1, 2, 3, \cdots$, where $\beta_j'(\lambda) = \frac{d}{d\lambda} \ \beta_j(\lambda)$.

In the following two lemmas we prove the existence and local Hölder continuity of the "trace maps".

<u>**Lemma 1.1**</u>

For each $\lambda > 0$ and $s > \frac{1}{2}$ there is a trace map, $T_0(\lambda)$, bounded from $L_s^2(\mathbf{R}^{n+1})$ into $L^2(S_c)$ such that for every $\varphi \in C_0^\infty(\mathbf{R}^{n+1})$

$$(T_0(\lambda)\varphi)(\omega) = \int \overline{\phi}_0(x, y, \lambda, \omega)\varphi(x, y)c_0^{-2}(y)\mu_0^{-1}(y)dxdy. \tag{1.67}$$

The function $\lambda \to T_0(\lambda)$ from $(0, \infty)$ into $\mathcal{B}(L_s^2(\mathbf{R}^{n+1}), \ L^2(S_c))$ is locally Hölder continuous with exponent γ, where $\gamma < 1$, and $\gamma < (s - 1/2)$, if $\lambda \neq \lambda_j, \quad j = 1, 2, 3, \cdots$, and $\gamma < 1/2, \ \gamma < (s - 1/2)$, if $\lambda = \lambda_j$, for some $j = 1, 2, 3, \cdots$.

<u>**Proof**</u>: For $\varphi(x, y) \in C_0^\infty(\mathbf{R}^{n+1})$ we define $T_0(\lambda)$ as given by (1.67). We prove below that it extends to a bounded operator that satisfies the requirements stated in the lemma. Note that

$$T_0(\lambda) = L_+(\lambda) + L_0(\lambda) + L_-(\lambda), \tag{1.68}$$

where

$$L_+(\lambda) = \chi_{S_+}(\lambda)T_0(\lambda),\ L_0(\lambda) = \chi_{S_0}(\lambda)T_0(\lambda),\ L_-(\lambda) = \chi_{S_-}(\lambda)T_0(\lambda), \tag{1.69}$$

where $\chi_{S_+}(\lambda)$, $\chi_{S_0}(\lambda)$, and $\chi_{S_-}(\lambda)$ are respectively the characteristic functions of S_+, S_0, and S_-. Moreover for $\alpha = +,\ 0,\ -$

$$L_\alpha(\lambda) = L_\alpha(\lambda)\ \chi_{(-\infty,0)}(y) + L_\alpha(\lambda)\ \chi_{(0,h)}(y) + L_\alpha(\lambda)\ \chi_{(h,\infty)}(y), \tag{1.70}$$

where $\chi_{(-\infty,0)}(y)$, $\chi_{(0,h)}(y)$, and $\chi_{(h,\infty)}(y)$ are respectively the characteristic functions of $(-\infty,0)$, $(0,h)$, and (h,∞).

a) The terms $L_\pm(\lambda)\chi_{(0,h)}(y)$.

For $\omega = (\omega_0, \overline{\omega}) \in S_\pm$ we denote (see (1.39) and (1.43) of Chapter 1)

$$\gamma_\pm = (c_\pm^{-2} - |\overline{\omega}|^2)^{1/2}. \tag{1.71}$$

Note that $\gamma_\pm = |\omega_0|$. Then by (1.54)

$$a_\pm = \left(\frac{\lambda^{1/2}}{\pi\mu_\pm}\right)^{1/2} \gamma_\pm^{1/2}\ b_\pm^{-1/2}, \tag{1.72}$$

where

$$b_\pm = \Big[\lambda\Big(\frac{\gamma_-}{\mu_-} + \frac{\gamma_+}{\mu_+}\Big)^2 (\cos(\lambda^{1/2}\gamma_h h))^2 + \Big(\frac{\lambda^{1/2}\gamma_h}{\mu_h} + \\ + \frac{\lambda^{1/2}\mu_h\gamma_+\gamma_-}{\mu_+\mu_-\gamma_h}\Big)^2 (\sin(\lambda^{1/2}\gamma_h h))^2\Big]. \tag{1.73}$$

Note that $b_\pm = 0$ if and only if $c_+ = c_-$, $\gamma_+ = \gamma_- = 0$, and $\lambda = \lambda_j$, for some $j = 1, 2, 3, \cdots$.

Moreover if $c_+ < c_-$ and $\lambda \in (0,\infty)$, or if $c_+ = c_-$ and $\lambda \in (0,\infty)$, $\lambda \neq \lambda_j$, $j = 1, 2, 3, \cdots$

$$\left|b_\pm^{-1/2}(\lambda,\omega)\right| \leq C, \tag{1.74}$$

where the constant C is uniform for λ in compact sets, and

$$\left|b_\pm^{-1/2}(\lambda_1,\omega) - b_\pm^{-1/2}(\lambda,\omega)\right| \leq C\ |\lambda_1 - \lambda|, \tag{1.75}$$

for all λ_1 in a neighborhood of λ, and some fixed constant C.

Furthermore if $c_+ = c_-$

$$\gamma_\pm \left|b_\pm^{-1/2}(\lambda,\omega)\right| \leq C, \quad \left|(\sin(\lambda^{1/2}\gamma_h h))\ b_\pm^{-1/2}(\lambda,\omega)\right| \leq C, \tag{1.76}$$

for $\lambda \in (0,\infty)$, where C is uniform for λ in compact sets, and

$$\gamma_\pm \left| b_\pm^{-1/2}(\lambda,\omega) - b_\pm^{-1/2}(\lambda_j,\omega)\right| \le \frac{C}{\gamma_\pm}\,|\lambda-\lambda_j|, \tag{1.77}$$

$$\left|\left(\sin(\lambda^{1/2}\gamma_h h)\right) b_\pm^{-1/2}(\lambda,\omega) - \left(\sin(\lambda_j^{1/2}\gamma_h h)\right) b_\pm^{-1/2}(\lambda_j,\omega)\right| \le \frac{C}{\gamma_\pm}\,|\lambda-\lambda_j|, \tag{1.78}$$

for λ in a neighborhood of λ_j, $\quad j=1,2,3,\cdots$. Then for all $0\le\delta\le 1$

$$\gamma_\pm \left| b_\pm^{-1/2}(\lambda,\omega) - b_\pm^{-1/2}(\lambda_j,\omega)\right| \le \frac{C}{\gamma_\pm^\delta}\,|\lambda-\lambda_j|^\delta, \tag{1.79}$$

$$\left|\left(\sin(\lambda^{1/2}\gamma_h h)\right) b_\pm^{-1/2}(\lambda,\omega) - \left(\sin(\lambda_j^{1/2}\gamma_h h)\right) b_\pm^{-1/2}(\lambda_j,\omega)\right| \le \frac{C}{\gamma_\pm^\delta}\,|\lambda-\lambda_j|^\delta, \tag{1.80}$$

for λ in a neighborhood of λ_j.

By (1.62), (1.64), and (1.67),

$$(L_\pm(\lambda)\chi_{(0,h)}\varphi)(\lambda,\omega) =$$

$$c_\pm^{1/2}\,\lambda^{n/4}\,\gamma_\pm^{1/2}\int_0^h \overline{\psi}_\pm(y,\lambda^{1/2}|\overline{\omega}|,\lambda)\hat{\varphi}(\lambda^{1/2}\overline{\omega},y)c_0^{-2}(y)\mu_0^{-1}(y)\,dx\,dy, \tag{1.81}$$

where $\hat{\varphi}(k,y)=(\mathcal{F}\varphi)(k,y)$ is the Fourier transform on the x variables of $\varphi(x,y)$ (see (1.11) in Chapter 1).

Let us denote by $T_S(\rho)$, $\rho>0$, the standard trace operator (see Reed and Simon 1975, Theorem IX.40, and Kuroda 1980, Theorems 4.2.1, 4.2.16, 4.3.1 and formula (7) page 4.30. Note that $T_S(\rho)$ corresponds to $\rho^{-\frac{n-1}{2}}$ times the trace operator in Kuroda's Theorem 4.2.16) from $H_s(\mathbf{R}^n)$, $s>1/2$, into $L^2(S_1^{n-1})$ such that

$$(T_S(\rho)\psi)(\nu)=\psi(\rho\,\nu),\quad \nu\in S_1^{n-1}, \tag{1.82}$$

for $\psi(k)\in C_0^\infty(\mathbf{R}^n)$.

Moreover the function $\rho\to T_S(\rho)$ from $(0,\infty)$ into $\mathcal{B}(H_s(\mathbf{R}^n),\ L^2(S_1^{n-1}))$ is locally Hölder continuous with exponent β satisfying $\beta<1$, $\beta<(s-1/2)$.

Furthermore for some constant C

$$\left\|T_S(\rho)\right\|_{\mathcal{B}\left(H_s(\mathbf{R}^n),\ L^2(S_1^{n-1})\right)} \le C\,\rho^{-\frac{(n-1)}{2}},\quad \rho\in(0,\infty), \tag{1.83}$$

and

$$\left\|\rho_1^{\frac{n-1}{2}}\, T_S(\rho_1) - \rho^{\frac{n-1}{2}}\, T_S(\rho)\right\|_{\mathcal{B}(H_s(\mathbf{R}^n),\ L^2(S_1^{n-1}))} \leq C\, |\rho - \rho_1|^{\beta}, \quad (1.84)$$

$\rho,\ \rho_1 \in (0,\infty),\ \beta < 1,\ \beta < (s-1/2)$.
Moreover $T_S(\rho)$ is surjective onto $H_{s-1/2}(S_1^{n-1})$ (see Lions and Magenes 1972, volume I, Theorem 9.4).

Then by (1.74), (1.76), and (1.81) (note that $\mathcal{F}$ sends $L_s^2(\mathbf{R}^n)$ into $H_s(\mathbf{R}^n)$)

$$\int_{S_{\pm}} \left|(L_{\pm}(\lambda)\chi_{(0,h)}\varphi)(\lambda,\omega)\right|^2 d\omega \leq C \int_0^h dy \int_{0<\rho<c_{\pm}^{-1}} \frac{d\rho}{\gamma_{\pm}(\rho)} +$$

$$+ \int_{|\nu|=1} d\nu\ \rho^{n-1} \left|\hat{\varphi}(\lambda^{1/2}\rho\nu, y)\right|^2 \left\|\gamma_{\pm}^{1/2}\psi_{\pm}\chi_{(0,h)}\right\|_{\mathcal{L}}^2 \leq C\, \|\varphi\|_{L_s^2(\mathbf{R}^{n+1})}. \quad (1.85)$$

It follows that $L_{\pm}(\lambda)\chi_{(0,h)}$ extends to a bounded operator from L_s^2 $(\mathbf{R}^{n+1})$ into $L^2(S_{\pm})$. Moreover by (1.79) if $c_+ = c_-$

$$\int_{S_{\pm}} \left|\left((L_{\pm}(\lambda) - L_{\pm}(\lambda_j))\chi_{(0,h)}\right)\varphi\right|^2 d\omega \leq C\left[|\lambda-\lambda_j|^2 + \right.$$

$$\left. + |\lambda-\lambda_j|^{2\gamma}\right] \|\varphi\|_{L_s^2(\mathbf{R}^{n+1})} + C \cdot \int_0^h dy \int_{0<\rho<c_{\pm}^{-1}} \frac{d\rho}{\gamma_{\pm}^{1+2\delta}(\rho)} \left|\lambda-\lambda_j\right|^{2\delta}$$

$$\int_{|\nu|=1} d\nu\ \rho^{n-1} \left|\hat{\varphi}(\lambda^{1/2}\rho\nu, y)\right|^2 \leq C\, |\lambda-\lambda_j|^{2\gamma} \|\varphi\|^2_{L_s^2(\mathbf{R}^{n+1})}, \quad (1.86)$$

$\gamma < 1/2,\ \gamma < (s-1/2)$. Then $L_{\pm}(\lambda)\chi_{(0,h)}$ is Hölder continuous at $\lambda = \lambda_j$ with exponent $\gamma < 1/2, \quad \gamma < (s-1/2)$.

When $\lambda \neq \lambda_j, \quad j = 1,2,3,\cdots$, or when $c_+ < c_-$, for all $\lambda \in (0,\infty)$, we prove as above (using (1.75) instead of (1.79)) that $L_{\pm}(\lambda)\chi_{(0,h)}$ is Hölder continuous at λ with exponent $\gamma < 1, \quad \gamma < (s-1/2)$.

b) The term $L_+(\lambda)\chi_{(h,\infty)}(y)$.

Note that $\chi_{S_+}\overline{\phi}_0\chi_{(h,\infty)}$ is a finite sum of terms of the form

$$d(\lambda,\omega)\, \frac{1}{(2\pi)^{n/2}}\, e^{-i\lambda^{1/2}\overline{\omega}\cdot x}\, e^{\mp i\lambda^{1/2}\omega_0 y}\, \chi_{(h,\infty)}(y), \quad (1.87)$$

(recall that $\omega_0 = \gamma_+$) where for $\lambda \in (0, \infty)$

$$|d(\lambda, \omega)| \leq C, \tag{1.88}$$

with C uniform for λ in compact sets. Moreover if $c_+ < c_-$, $\lambda \in (0, \infty)$, or $c_+ = c_-$ and $\lambda \in (0, \infty)$, $\lambda \neq \lambda_j$, $j = 1, 2, 3, \cdots$,

$$|d(\lambda_1, \omega) - d(\lambda, \omega)| \leq C \, |\lambda_1 - \lambda|, \tag{1.89}$$

for all λ_1 in a neighborhood of λ. If $c_+ = c_-$ for $j = 1, 2, 3, \cdots$,

$$|d(\lambda, \omega) - d(\lambda_j, \omega)| \leq \frac{C}{\gamma_+} \, |\lambda - \lambda_j|, \tag{1.90}$$

for all λ in a neighborhood of λ_j.

Then for all $0 \leq \delta \leq 1$

$$|d(\lambda, \omega) - d(\lambda_j, \omega)| \leq \frac{C}{\gamma_+^{\delta}} \, |\lambda - \lambda_j|^{\delta}. \tag{1.91}$$

Each term of the type (1.87) gives a contribution to $L_+(\lambda)\chi_{(h,\infty)}$ of the form

$$(D(\lambda)\varphi)(\omega) = d(\lambda, \omega)(T_S(\lambda^{1/2})(\mathcal{F}_1\chi_{(h,\infty)}(\pm y)c_0^{-2}(\pm y)\mu_0^{-1}(\pm y)\varphi(x, \pm y)), \tag{1.92}$$

where $\mathcal{F}_1$ denotes the unitary operator of Fourier transform on L^2 $(\mathbf{R}^{n+1})$. Then since $T_S(\lambda^{1/2})$ is bounded, $D(\lambda)$ is a bounded operator from $L^2_s(\mathbf{R}^{n+1})$ into $L^2(S_+)$. Moreover if $c_+ < c_-$, $\lambda \in (0, \infty)$, or if $c_+ = c_-$ and $\lambda \neq \lambda_j$, $j = 1, 2, 3, \cdots$, it follows from (1.89) that $D(\lambda)$ is locally Hölder continuous with exponent $\gamma < 1$, $\gamma < (s - 1/2)$.

Suppose that $c_+ = c_-$, and $\lambda = \lambda_j$, for some $j = 1, 2, 3, \cdots$.

Let $f(\omega) \in C^\infty(S^n_{\frac{1}{c_+}})$ satisfy $f(\omega) \equiv 1$ for ω in a neighborhood of $\omega_0 = 0$. Then

$$D(\lambda) = f(\omega) \, D(\lambda) + (1 - f(\omega)) \, D(\lambda). \tag{1.93}$$

Since

$$|(1 - f(\omega)) \, (d(\lambda, \omega) - d(\lambda_j, \omega))| \leq C \, |\lambda - \lambda_j|, \tag{1.94}$$

for λ in a neighborhood of λ_j, $(1 - f(\omega)) \, D(\lambda)$ is Hölder continuous at $\lambda = \lambda_j$, with exponent $\gamma < 1$, $\gamma < (s - 1/2)$.

By (1.90)

$$\int_{S_+} f^2(\omega) \, |(D(\lambda) - D(\lambda_j))\varphi|^2 d\omega \leq C \, |\lambda - \lambda_j|^{2\gamma} \, \left\|\varphi\right\|_{L^2_s(\mathbf{R}^{n+1})} +$$

$$+ \, C \, |\lambda - \lambda_j|^{2\delta} \int_{S_+} f^2(\omega) \, \frac{d\omega}{\omega_0^{2\delta}} \, |T_S(\lambda_j^{1/2})\mathcal{F}_1(c_0^{-2}(\pm y)\mu_0(\pm y)\varphi(x, \pm y))|^2. \tag{1.95}$$

Let $g(x) \in C_0^\infty(\mathbf{R})$ with $g(x) \equiv 1$ in a neighborhood of $x = 0$. Then by Sobolev theorem (see Adams 1975)

$$\left\| g \frac{\varphi}{|x|^\delta} \right\|_{L^2(\mathbf{R})} \leq \left\| \varphi \right\|_\infty \left\| \frac{g}{|x|^\delta} \right\|_{L^2(\mathbf{R})} \leq C \left\| \varphi \right\|_{H_s(\mathbf{R})}, \tag{1.96}$$

for $0 \leq \delta < 1/2$, and $s > 1/2$. Since the function $(1 - g(x)) \frac{1}{|x|^\delta}$ is bounded, it follows that $\frac{1}{|x|^\delta} \in \mathcal{B}(H_s(\mathbf{R}), \quad L^2(\mathbf{R}))$, for $s > 1/2$, and $0 \leq \delta < 1/2$. It follows by interpolation (see Lions and Magenes 1972, and Reed and Simon 1975) that $\frac{1}{|x|^\delta} \in \mathcal{B}(H_s(\mathbf{R}), \quad L^2(\mathbf{R}))$ for $\delta < s, \quad 0 \leq \delta < 1/2$.

Then by taking ω_0 as one of the coordinates in S_+ and standard localization it follows from (1.95) that $\Big($recall that $T_S(\rho) \in \mathcal{B}(H_s(\mathbf{R}^{n+1}),$ $H_{s-1/2}(S_1^n))\Big)$ $f(\omega)D(\lambda)$ is Hölder continuous at $\lambda = \lambda_j$ with exponent $\gamma < 1/2, \quad \gamma < (s - 1/2)$.

c) The term $L_-(\lambda)\chi_{(-\infty,0)}(y)$.

As in case (b) above we prove that $L_-(\lambda)\chi_{(-\infty,0)}$ extends to a bounded operator from $L_s^2(\mathbf{R}^{n+1})$ into $L^2(S_-)$ and that the function $\lambda \to L_-(\lambda)$ $\chi_{(-\infty,0)}(y)$ is Hölder continuous with the same exponent γ as in cases (a) and (b) above.

d) The term $L_+(\lambda)\chi_{(-\infty,0)}(y)$.

If $c_- = c_+$ we prove as in case (b) above that $L_+(\lambda)\chi_{(-\infty,0)}$ extends to a bounded operator that is Hölder continuous with exponent as in case (b).

Suppose that $c_+ < c_-$ and $\lambda \in (0, \infty)$. Let $f(\rho) \in C_0^\infty((0,\infty))$ satisfy $f(\rho) = 1$, for ρ in a neighborhood of $\rho = \lambda^{1/2}$. For $\varphi \in C_0^\infty(\mathbf{R}^{n+1})$ denote

$$(D\varphi)(\rho, \omega) = \frac{f(\rho)}{(2\pi)^{n/2}} \int e^{-i\rho\overline{\omega}\cdot x}\, e^{-i\rho\gamma_- y}\, c_0^{-2}(y)\mu_0^{-1}(y)\varphi(x,y)dxdy. \tag{1.97}$$

Then with $\xi = \rho\omega$

$$\int\limits_{(0,\infty)\times S_+} |D\varphi|^2 d\rho d\omega \leq C \int\limits_{\mathbf{R}^n} d\overline{\xi} \int\limits_{a|\overline{\xi}|}^{\infty} d\xi_0\, |(\mathcal{F}_1\, c_0^{-2}\mu_0^{-1}\varphi)(\xi_0, \overline{\xi})|^2 \leq$$

$$\leq C \int\limits_{\mathbf{R}^n} d\overline{\xi} \int\limits_0^\infty dz\, |(\mathcal{F}_1(c_0^{-2}\mu_0^{-1}\varphi))(z, \overline{\xi})|^2, \tag{1.98}$$

where $z = |\xi| c_+ \gamma_- = \left(\xi_0^2\, \frac{c_+^2}{c_-^2} - \left(1 - \frac{c_+^2}{c_-^2}\right)|\overline{\xi}|^2 \right)^{1/2}$.

Then since $\mathcal{F}_1$ is unitary on $L^2(\mathbf{R}^{n+1})$

$$\left\| D\varphi \right\|_{L^2\left((0,\infty),\ L^2(S_+)\right)} \leq C \left\| \varphi \right\|_{L^2(\mathbf{R}^{n+1})}, \tag{1.99}$$

and $D \in \mathcal{B}\Big(L^2(\mathbf{R}^{n+1}), L^2((0,\infty), L^2(S_+))\Big)$. We similarly prove that $D \in \mathcal{B}\Big(L_1^2(\mathbf{R}^{n+1}),\ H_1((0,\infty),\ L^2(S_+))\Big)$, and then by interpolation (see Lions and Magenes 1972, and Reed and Simon 1975) that $D \in \mathcal{B}\Big(L_s^2(\mathbf{R}^{n+1}), H_s((0,\infty), L^2(S_+))\Big)$, $\quad 0 \leq s \leq 1$.

Let us denote by $T_S(\rho)$ the standard trace operator from $H_s((0,\infty)$, $L^2(S_+))$ into $L^2(S_+)$ (see Lions and Magenes 1972, and Kuroda 1980, Theorem 16 in Section 2 of Chapter 4). Then

$$(L_+(\lambda)\chi_{(-\infty,0)}\varphi)(\omega) = c_+\ \lambda^{n/4}\ \gamma_+^{1/2}\ a_+(T_S(\lambda^{1/2})D\varphi)(\omega), \tag{1.100}$$

and it follows by (1.74) and (1.75) that $L_+(\lambda)\chi_{(-\infty,0)}$ extends to a bounded operator that is locally Hölder continuous with exponent $\gamma < 1$, $\gamma < (s-\frac{1}{2})$.

c) The term $L_-(\lambda)\chi_{(h,\infty)}$.

We prove as in case (d) above that $L_-(\lambda)\chi_{(h,\infty)}$ extends to a bounded operator from $L_s^2(\mathbf{R}^{n+1})$ into $L^2(S_-)$ that is locally Hölder continuous with exponent as in case (d).

f) The terms $L_0(\lambda)\chi_{(h,\infty)}$, and $L_0(\lambda)\chi_{(h,\infty)}$.

Note that (see (1.60))

$$a_0(\lambda,\omega) = \left(\frac{\lambda^{1/2}}{\pi\mu_+}\right)^{1/2}\ \gamma_+^{1/2}\ b_0^{-1/2}, \tag{1.101}$$

where

$$b_0 = \left[\lambda\ \frac{\gamma_+^2}{\mu_+^2}\ \left(\cos(\lambda^{1/2}\ \gamma_h h) + \frac{\mu_h\ \gamma_-^{\sim}}{\mu_-\ \gamma_h}\ \sin(\lambda^{1/2}\ \gamma_h h)\right)^2 + \right.$$
$$\left. + \left(\lambda^{1/2}\ \frac{\gamma_-^{\sim}}{\mu_-}\ \cos(\lambda^{1/2}\gamma_h h) - \frac{\lambda^{1/2}\ \gamma_h}{\mu_h}\ \sin(\lambda^{1/2}\ \gamma_h h)\right)^2\right], \tag{1.102}$$

where

$$\gamma_-^{\sim} = (|\overline{\omega}|^2 - c_-^{-2})^{1/2}. \tag{1.103}$$

Moreover $b_0 = 0$ if and only if $\gamma_+ = 0$, and $\lambda = \lambda_j$, for some $j = 1, 2, 3, \cdots$.

If $\lambda \in (0,\infty)$, $\quad \lambda \neq \lambda_j$, $\quad j = 1, 2, \cdots$

$$|b_0^{-1/2}(\lambda,\omega)| \leq C, \tag{1.104}$$

where C is uniform for λ in compact sets, and

$$|b_0^{-1/2}(\lambda_1,\omega) - b_0^{-1/2}(\lambda,\omega)| \leq C\,|\lambda_1 - \lambda|, \tag{1.105}$$

for all λ_1 in a neighborhood of λ, and some fixed constant C.

Moreover for $\lambda \in (0,\infty)$

$$|\gamma_+\, b_0^{-1/2}(\lambda,\omega)| \leq C, \; \Big|\Big(\frac{\gamma_-^{\sim}}{\mu_-}\,\cos(\lambda^{1/2}\,\gamma_h h) - \\ - \frac{\gamma_h}{\mu_h}\,\sin(\lambda^{1/2}\,\gamma_h h)\Big) b_0^{-1/2}(\lambda,\omega)\Big| \leq C, \tag{1.106}$$

with C uniform for λ in compact sets of $(0,\infty)$, and for $j = 1,2,3,\cdots$,

$$\Big|\gamma_+\,\Big(b_0^{-1/2}(\lambda,\omega) - b_0^{-1/2}(\lambda_j,\omega)\Big)\Big| \leq \frac{C}{\gamma_+}\,|\lambda - \lambda_j|, \tag{1.107}$$

$$\Big|\lambda^{1/2}\,\Big(\frac{\gamma_-^{\sim}}{\mu_h}\,\cos(\lambda^{1/2}\,\gamma_h h) - \frac{\gamma_h}{\mu_h}\,\sin(\lambda^{1/2}\,\gamma_h h)\Big)\; b_0^{-1/2}(\lambda,\omega) - \\ - \lambda_j^{1/2}\,\Big(\frac{\gamma_-^{\sim}}{\mu_h}\,\cos(\lambda_j^{1/2}\,\gamma_h h) - \frac{\gamma_h}{\mu_h}\,\sin(\lambda_j^{1/2}\,\gamma_h h)\Big)\; b_0^{-1/2}(\lambda_j,\omega)\Big| \leq \\ \leq \frac{C}{\gamma_+}\,|\lambda - \lambda_j|. \tag{1.108}$$

Then as in cases (a) and (b) above we prove that the terms $L_0(\lambda)\chi_{(0,\infty)}$ and $L_0(\lambda)\chi_{(h,\infty)}$ extend to bounded operators that are locally Hölder continuous with exponent γ that satisfies $\gamma < 1$, $\gamma < (s - 1/2)$ if $\lambda \neq \lambda_j$, $j = 1,2,3,\cdots$, and $\gamma < 1/2$, $\gamma < (s - 1/2)$ if $\lambda = \lambda_j$ for some $j = 1,2,\cdots$.

g) The term $L_0(\lambda)\chi_{(-\infty,0)}(y)$.

Note that

$$\Big\|\gamma_+^{1/2}\;a_0\;e^{\lambda^{1/2}\;\gamma_-^{\sim} y}\;\chi_{(-\infty,0)}\Big\|_{L^2(\mathbf{R})} \leq \frac{C}{(\gamma_-^{\sim})^{1/2}}. \tag{1.109}$$

Then as in case (a) above we prove that $L_0(\lambda)\chi_{(-\infty,0)}(y)$ extends to a bounded operator from $L_s^2(\mathbf{R}^{n+1})$ into $L^2(S_0)$. Note that since $c_+ < c_-$, γ_+ and γ_- are not zero simultaneously.

Moreover if $\lambda \neq \lambda_j$, $j = 1,2,3,\cdots$,

$$\Big\|\gamma_+^{1/2}(a_0(\lambda_1,\omega)\;e^{\lambda_1^{1/2}\;\gamma_-^{\sim} y} - a_0(\lambda,\omega)\;e^{\lambda^{1/2}\;\gamma_-^{\sim} y})\chi_{(-\infty,0)}\Big\|_{L^2(\mathbf{R})} \leq \\ \leq \frac{C}{(\gamma_-^{\sim})^{1/2}}\,|\lambda_1 - \lambda|, \tag{1.110}$$

for all λ_1 in a neighborhood of λ, and for $j = 1, 2, 3, \cdots$,

$$\left\| \gamma_+^{1/2} (a_0(\lambda, \omega)\, e^{\lambda^{1/2}\, \gamma_-^{\sim} y} - a_0(\lambda_j, \omega)\, e^{\lambda_j^{1/2}\, \gamma_-^{\sim} y}) \chi_{(-\infty,0)} \right\|_{L^2(\mathbf{R})} \leq$$

$$\leq C\, \frac{1}{\gamma_+}\, \frac{1}{(\gamma_-^{\sim})^{1/2}}\, |\lambda - \lambda_j|. \quad (1.111)$$

Then as in case (a) above we prove that $L_0(\lambda)\chi_{(-\infty,0)}(y)$ is locally Hölder continuous with exponent γ that satisfies $\gamma < 1$, $\gamma < (s - 1/2)$ if $\lambda \neq \lambda_j$, $j = 1, 2, 3, \cdots$, and $\gamma < 1/2$, $\gamma < (s - 1/2)$, if $\lambda = \lambda_j$, for some $j = 1, 2, 3, \cdots$.

Q. E. D.

<u>**Lemma 1.2**</u>

For each $j = 1, 2, 3, \cdots$, $\lambda \in (\lambda_j, \infty)$ and $s > 1/2$, there is a trace map, $T_j(\lambda)$, bounded from $L^2_s(\mathbf{R}^{n+1})$ into $L^2(S_1^{n-1})$ such that for every $\varphi \in C_0^\infty(\mathbf{R}^{n+1})$, $\lambda > \lambda_j$,

$$(T_j(\lambda)\varphi)(\nu) = \int \overline{\phi}_j(x, y, \lambda, \nu)\varphi(x, y) c_0^{-2}(y) \mu_0^{-1}(y)\, dx dy. \quad (1.112)$$

Moreover the functions $\lambda \to T_j(\lambda)$ from (λ_j, ∞) into $\mathcal{B}(L^2_s(\mathbf{R}^{n+1}),\ L^2(S_1^{n-1}))$ are locally Hölder continuous with exponent $\gamma < 1$, $\gamma < (s - 1/2)$. Furthermore if we extend $T_j(\lambda)$ to $\lambda = \lambda_j$ as $T_j(\lambda_j) = 0$, for $j = 2, 3, 4, \cdots$, and if $c_+ < c_-$, $T_1(\lambda_1) = 0$, then $T_j(\lambda)$ is Hölder continuous at $\lambda = \lambda_j$, with exponent $\gamma \leq 1/2$, $\gamma < (s - 1/2)$.

<u>**Proof**</u>: Take $\lambda > \lambda_j$, and for $\varphi(x, y) \in C_0^\infty(\mathbf{R}^{n+1})$ define $T_j(\lambda)$ by (1.112). Then denoting $\hat{\varphi}(k, y) = (\mathcal{F}\varphi)(k, y)$,

$$(T_j(\lambda)\varphi)(\nu) = (\beta_j(\lambda))^{\frac{n-1}{2}}\ (\beta_j^{\prime}(\lambda))^{1/2} \int \psi_j(y, \beta_j)\hat{\varphi}(\beta_j \nu, y)$$

$$c_0^{-2}(y) \mu_0^{-1}(y)\, dx dy. \quad (1.113)$$

As in the proof of Lemma 1.1 we denote by $T_S(\rho)$ the standard trace operator from $H_s(\mathbf{R}^n)$ into $L^2(S_1^{n-1})$. Then for any $\epsilon \geq 0$, $\epsilon' > 0$, such that $\frac{1}{2} + \epsilon + \epsilon' \leq s$,

$$\left\|T_j(\lambda)\varphi\right\|^2_{L^2(S_1^{n-1})} \leq$$

$$\leq C\left\|(1+y^2)^{-\epsilon/2}\,\psi_j\right\|^2_{L^2(\mathbf{R})}\int dy(1+y^2)^{\epsilon}\,\left\|T(\beta_j)\hat{\varphi}\right\|^2_{L^2(S_1^{n-1})} \leq$$

$$\leq C\,\left\|(1+y^2)^{-\epsilon/2}\,\psi_j\right\|^2_{\mathcal{L}}\,\left\|\varphi\right\|^2_{L^2_s(\mathbf{R}^{n+1})}. \tag{1.114}$$

Then $T_j(\lambda)$ extends to a bounded operators from $L^2_s(\mathbf{R}^{n+1})$ into $L^2(S_1^{n-1})$.

It follows by (1.40), (1.41) that for $\lambda > \lambda_j$

$$\left\|\psi_j(y,\beta_j(\lambda_1)) - \psi_j(y,\beta_j(\lambda))\right\|_{\mathcal{L}} \leq C\,|\lambda_1 - \lambda|, \tag{1.115}$$

for a fixed constant C and all λ_1 in a neighborhood of λ. Then by the Hölder continuity of the trace operators $T_S(\rho)$ we prove, estimating as in (1.114) with $\epsilon = 0$, that $T_j(\lambda)$ is Hölder continuous at $\lambda > \lambda_j$ with exponent $\gamma < 1$, $\gamma < (s - 1/2)$.

Moreover by (1.42)

$$\int_0^h (1+y^2)^{-\epsilon}\,|\psi_j(y,\beta_j(\lambda))|^2\,dy \leq C\,\lambda^{\sim}_{+,j}. \tag{1.116}$$

Furthermore

$$\int_h^\infty e^{-2\lambda^{\sim}_{+,j}y}\,dy \leq \frac{C}{2\lambda^{\sim}_{+,j}}, \tag{1.117}$$

$$\int_h^\infty \frac{e^{-2\lambda^{\sim}_{+,j}y}}{(1+y^2)^{\frac{1+\Delta}{2}}}\,dy \leq C, \tag{1.118}$$

for $\Delta > 0$.

Then by interpolation,

$$\int_h^\infty \frac{e^{-2\lambda^{\sim}_{+,j}y}}{(1+y^2)^{\epsilon}}\,dy \leq C\,(\lambda^{\sim}_{+,j})^{\frac{2\,\epsilon}{1+\Delta}-1}, \tag{1.119}$$

for $0 \leq \epsilon \leq \frac{1+\Delta}{2}$. It follows that

$$\int_h^\infty \frac{1}{(1+y^2)^{\epsilon}}\,|\psi_j(y,\beta_j(\lambda))|^2\,dy \leq C\,(\lambda^{\sim}_{+,j})^{\frac{2\epsilon}{1+\Delta}}. \tag{1.120}$$

If $c_+ < c_-$ we have that

$$\int\limits_{-\infty}^{0} \frac{1}{(1+y^2)^{\epsilon}} \, |\psi_j(y, \beta_j(\lambda))|^2 \, dy \leq C \, \lambda^{\sim}_{+,j}. \tag{1.121}$$

If $c_+ = c_-$ we estimate as in (1.120) to obtain

$$\int\limits_{-\infty}^{0} \frac{1}{(1+y^2)^{\epsilon}} \, |\psi_j(y, \beta_j(\lambda))|^2 \, dy \leq C \, (\lambda^{\sim}_{+,j})^{\frac{2\epsilon}{1+\Delta}}. \tag{1.122}$$

By (1.117), (1.120), and (1.122)

$$\left\| (1+y^2)^{-\epsilon/2} \, \psi_j(y, \beta_j(\lambda)) \right\|_{L^2(\mathbf{R})} \leq C \, (\lambda^{\sim}_{+,j})^{\gamma}, \quad \gamma \leq \frac{\epsilon}{1+\Delta}.$$

Then for $j = 2, 3, \cdots$, (and if $c_+ < c_-$ also for $j = 1$) by (1.114)

$$\left\| T_j(\lambda)\varphi \right\|_{L^2(S_1^{n-1})} \leq C \, |\lambda - \lambda_j|^{\gamma} \, \left\| \varphi \right\|_{L^2_s(\mathbf{R}^{n+1})}, \tag{1.123}$$

with $\gamma \leq 1/2, \quad \gamma < (s - 1/2)$, where we used (1.38) and (1.39).

Q. E. D.

Appendix 2

In this appendix we state the results on eigenfunctions expansion and the limiting absorption principle of an operator that is related to the unperturbed acoustic propagator.

We denote by $\mathcal{N}_{\mu_0}$ the Hilbert space consisting of all the complex valued, measurable, square integrable functions on $\mathbf{R}$ with scalar product

$$\left(\varphi, \psi\right)_{\mathcal{N}_{\mu_0}} = \int \varphi(y)\overline{\psi}(y) \frac{1}{\mu_0(y)} dy, \tag{1.1}$$

where

$$\mu_0(y) = \begin{cases} \mu_+, & y \geq h, \\ \mu_h, & h > y \geq 0, \\ \mu_-, & y < 0, \end{cases} \tag{1.2}$$

for some positive constants μ_+, μ_h, μ_-, and h.

Let $f(\,\cdot\,,\,\cdot\,)$ be the following quadratic form with domain $H_1(\mathbf{R})$

$$f(\varphi, \psi) = \int_{\mathbf{R}} \left(\frac{d}{dy}\varphi \frac{d}{dy} \overline{\psi} + q(y)\varphi\overline{\psi}\right) \frac{dy}{\mu_0(y)}, \tag{1.3}$$

where

$$q(y) = \begin{cases} 0, & y \geq h, \\ -q_h, & h > y \geq 0, \\ q_-, & y < 0, \end{cases} \tag{1.4}$$

with $q_h > 0$, and $q_- \geq 0$.

Let h be the selfadjoint bounded below operator associated with the quadratic form $f(\,\cdot\,,\,\cdot\,)$ (see Kato 1976, Chapter 6). Then

$$D(h) = \left\{\varphi \in H_1(\mathbf{R}) \,:\, \frac{d}{dy} \frac{1}{\mu_0} \varphi \in \mathcal{N}_{\mu_0}\right\}, \tag{1.5}$$

and for $\varphi \in D(h)$

$$h\varphi = -\mu_0 \frac{d}{dy} \frac{1}{\mu_0} \frac{d}{dy} \varphi + q\varphi. \tag{1.6}$$

The following facts concerning the eigenvalues, eigenfunctions and generalized eigenfunctions of h are established by explicit calculation.

h has no positive or zero eigenvalues. It has a finite number, Q, of negative eigenvalues, λ_j, $\quad -q_h < \lambda_j < 0, \quad 1 \leq j \leq Q$, of multiplicity one, where $\lambda_1 < \lambda_2 < \lambda_3 < \cdots < \lambda_Q$, with associated eigenfunction

$$\psi_j = a_j \begin{cases} e^{-\rho_j^+ y}, & y \geq h, \\ b_j \cos(\rho_j y) + c_j \sin(\rho_j y), & 0 \leq y \leq h, \\ d_j \, e^{\rho_j^+ y}, & y \leq 0, \end{cases} \tag{1.7}$$

for some real constants a_j, b_j, c_j, and d_j, and where

$$\left\| \psi_j \right\|_{\mathcal{N}_{\mu_0}} = 1. \tag{1.8}$$

Moreover $\rho_j^+ = (-\lambda_j)^{1/2}$, $\rho_j = (q_h + \lambda_j)^{1/2}$, and $\rho_j^- = (q_- - \lambda_j)^{1/2}$.

For $\rho > (q_-)^{1/2}$, h has two generalized eigenfunctions $\psi_\pm(\rho_-, y)$ that satisfy

$$\left(-\mu_0 \frac{d}{dy} \frac{1}{\mu_0} \frac{d}{dy} + q \right) \psi_\pm = \rho^2 \, \psi_\pm, \tag{1.9}$$

where $\rho_- = (\rho^2 - q_-)^{1/2}$, and

$$\psi_+(\rho_-, y) = \left(\frac{\mu_+ \, \rho_-}{2\pi\rho} \right)^{1/2} \begin{cases} \alpha_+ \, e^{i\rho y} + e^{-i\rho y}, & y \geq h, \\ \beta_+ \, e^{i\rho_0 y} + \gamma_+ \, e^{-i\rho_0 y}, & h \geq y \geq 0, \\ \delta_+ \, e^{-i\rho_- y}, & y \leq 0, \end{cases} \tag{1.10}$$

$$\psi_-(\rho_-, y) = \left(\frac{\mu_-}{2\pi} \right)^{1/2} \begin{cases} \alpha_- \, e^{i\rho y}, & y \geq h, \\ \beta_- \, e^{i\rho_0 y} + \gamma_- \, e^{-i\rho_0 y}, & h \geq y \geq 0, \\ e^{i\rho_- y} + \delta_- \, e^{-i\rho_- y}, & y \leq 0, \end{cases} \tag{1.11}$$

where $\rho_0 = (\rho^2 + q_h)^{1/2}$, and

$$\alpha_+ = -\frac{P_2}{P_1}, \; \beta_+ = -\frac{P_2 \, P_3}{P_1} + P_4, \; \gamma_+ = -\frac{P_2 \overline{P}_4}{P_1} + \overline{P}_3, \; \delta_+ = \frac{\mu_- \rho}{\mu_+ \, \rho_-} \frac{1}{P_1} \tag{1.12}$$

$$\alpha_- = \frac{1}{P_1}, \; \beta_- = \frac{P_3}{P_1}, \; \gamma_- = \frac{\overline{P}_4}{P_1}, \; \delta_- = \frac{\overline{P}_2}{P_1}, \tag{1.13}$$

moreover

$$P_1 = \frac{e^{i\rho h}}{2}\left[\frac{\mu_- \rho + \mu_+ \rho_-}{\mu_+ \rho_-}\cos(\rho_0 h) - \frac{i(\mu_h^2 \rho\, \rho_- + \mu_+ \mu_- \rho_0^2)}{\mu_+ \mu_h\, \rho_0 \rho_-}\sin(\rho_0 h)\right], \tag{1.14}$$

$$P_2 = \frac{e^{-i\rho h}}{2}\left[\frac{\mu_+ \rho_- - \mu_- \rho}{\mu_+ \rho_-}\cos(\rho_0 h) + \frac{i(\mu_h^2 \rho\, \rho_- - \mu_+ \mu_- \rho_0^2)}{\mu_+ \mu_h\, \rho_0 \rho_-}\sin(\rho_0 h)\right]. \tag{1.15}$$

$$P_3 = \frac{\mu_+ \rho_0 + \mu_h \rho}{2\, \mu_+\, \rho_0}\; e^{i(\rho-\rho_0)h}, \tag{1.116}$$

$$P_4 = \frac{\mu_+ \rho_0 - \mu_h \rho}{2\, \mu_+\, \rho_0}\; e^{-i(\rho+\rho_0)h}. \tag{1.117}$$

For $0 < \rho < \sqrt{q_-}$, h has one generalized eigenfunction $\psi_0(\rho_-^{\sim}, y)$, that satisfies

$$\Big(-\mu_0 \frac{d}{dy}\frac{1}{\mu_0}\frac{d}{dy} + q\Big)\psi_0 = \rho^2\, \psi_0, \tag{1.18}$$

where $\rho_-^{\sim} = (q_- - \rho^2)^{1/2}$, and

$$\psi_0(\rho_-^{\sim}, y) =$$

$$= \left(\frac{2\mu_+ \rho_-^{\sim}}{\pi\, \rho}\right)^{1/2}\left(\alpha_0^2 + \beta_0^2\right)^{-1/2}\begin{cases} \alpha_0 \cos(\rho y) + \beta_0 \sin(\rho y), & y \geq h, \\ \cos(\rho_0 y) + \frac{\mu_h \rho_-^{\sim}}{\mu_- \rho_0}\sin(\rho_0 y), & h \geq y \geq 0, \\ e^{\rho_-^{\sim} y}, & y \leq 0, \end{cases} \tag{1.19}$$

where $\alpha_0 = \gamma_0 \cos(\rho h) - \frac{\delta_0}{\rho}\sin(\rho h)$, $\beta_0 = \frac{1}{\sin(\rho h)}\,(\gamma_0 - \alpha_0 \cos(\rho h))$, if $\sin(\rho h) \neq 0$, and $\beta_0 = \frac{\delta_0}{\rho}$, if $\sin(\rho h) = 0$, with $\gamma_0 = \cos(\rho_0 h)$ $+ \frac{\mu_h}{\mu_-}\frac{\rho_-^{\sim}}{\rho_0}\sin(\rho_0 h)$, $\delta_0 = \frac{-\mu_+}{\mu_h}\rho_0 \sin(\rho_0 h) + \frac{\mu_+}{\mu_-}\rho_-^{\sim}\cos(\rho_0 h)$.

The following results follow from the theory of Weyl, Kodaira, and Titchmarsh (see Dunford and Schwartz 1963, Chapter XIII, and Wilcox 1984, Appendix 1, and the proof of Theorems 6.1 and 6.4 in Chapter 3).

For every $\varphi \in \mathcal{N}_{\mu_0}$ the following limits

$$(F_+\varphi)(\rho_-) = s - \lim_{N\to\infty}\int_{-N}^{N} \overline{\psi}_+(\rho_-, y)\varphi(y)\,\frac{dy}{\mu_0(y)}, \tag{1.20}$$

$$(F_-\varphi)(\rho_-) = s - \lim_{N\to\infty} \int_{-N}^{N} \overline{\psi}_-(\rho_-, y)\varphi(y)\,\frac{dy}{\mu_0(y)}, \tag{1.21}$$

$$(F_0\varphi)(\rho_-^{\sim}) = s - \lim_{N\to\infty} \int_{-N}^{N} \overline{\psi}_0(\rho_-^{\sim}, y)\varphi(y)\,\frac{dy}{\mu_0(y)}, \tag{1.22}$$

exist respectively in the strong topology in $L^2(0,\infty)$, $L^2(0,\infty)$, and $L^2(0, \sqrt{q_-})$. The operators F_+, F_-, and F_0, defined respectively by (1.20) to (1.22) are partially isometric from $\mathcal{N}_{\mu_0}$ onto $L^2(0,\infty)$, $L^2(0,\infty)$, and $L^2(0,\sqrt{q_-})$. If we furthermore define the operators

$$F_j\varphi = (\varphi, \psi_j), \quad 1 \le j \le Q, \tag{1.23}$$

the operator F_h defined as

$$F_h\varphi = F_+\varphi \,\oplus\, F_-\varphi \bigoplus_{j=0}^{Q} F_j\varphi, \tag{1.24}$$

is unitary from $\mathcal{N}_{\mu_0}$ onto

$$L^2(0,\infty) \,\oplus\, L^2(0,\infty) \,\oplus\, L^2(0,\sqrt{q_-}) \bigoplus_{j=1}^{Q} \mathbf{C}. \tag{1.25}$$

Moreover for any $\varphi \in D(h)$

$$F_h(h\varphi) = (\rho_-^2 + q_-)F_+\varphi \oplus (\rho_-^2 + q_-)F_-\varphi \oplus (q_- - \rho_-^{\sim 2})F_0\varphi \bigoplus_{j=1}^{Q} \lambda_j F_j\varphi. \tag{1.26}$$

Let us denote by

$$\Delta_x = \sum_{i=1}^{n} \frac{\partial^2}{\partial x_i^2}, \tag{1.27}$$

the selfadjoint realization of the Laplacian in $L^2(\mathbf{R}^n)$, with domain$H_2(\mathbf{R}^n)$.

Let us denote by $\mathcal{H}_{\mu_0}$ the Hilbert space

$$\mathcal{H}_{\mu_0} = L^2(\mathbf{R}^{n+1}) \otimes \mathcal{N}_{\mu_0}. \tag{1.28}$$

We define

$$B = -\Delta_x \otimes I_{\mathcal{N}_{\mu_0}} + I_{L^2(\mathbf{R}^n)} \otimes h, \tag{1.29}$$

where $I_{\mathcal{N}_{\mu_0}}$, and $I_{L^2(\mathbf{R}^n)}$ are respectively the identity operator on $\mathcal{N}_{\mu_0}$ and $L^2(\mathbf{R}^n)$. B is a selfadjoint operator in $\mathcal{H}_{\mu_0}$.

The generalized eigenfunctions of B are obtained from those of h as follows:

$$\psi^{\pm}(k,x,y) = \frac{1}{(2\pi)^{n/2}}\, e^{i\overline{k}\cdot x}\, \psi_{\pm}(k_0,y), \tag{1.30}$$

where $k = (k_0, \overline{k}) \in \mathbf{R}^{n+1}$, with

$$\mathbf{R}^{n+1}_{+} = \{k \in \mathbf{R}^{n+1} \ : \ k_0 > 0\}, \tag{1.31}$$

$$\psi^{0}(k,x,y) = \frac{1}{(2\pi)^{n/2}}\, e^{i\overline{k}\cdot x}\, \psi_{0}(k_0,y), \tag{1.32}$$

for $k = (k_0, \overline{k}) \in \Omega_0$, where

$$\Omega_0 = \{k \in \mathbf{R}^{n+1} \ : \ 0 < k_0 < \sqrt{q_-}\ \}, \tag{1.33}$$

and for $j = 1, 2, 3, \cdots, Q$,

$$\psi^{j}(\overline{k},x,y) = \frac{1}{(2\pi)^{n/2}}\, e^{i\overline{k}\cdot x}\, \psi_{j}(y), \tag{1.34}$$

for $\overline{k} \in \mathbf{R}^n$.

Then since the Fourier transform is a unitary operator that gives a spectral representation for the Laplacian:
For every $\varphi \in \mathcal{H}_{\mu_0}$ the following limits

$$(F^{\pm}\varphi)(k) = s - \lim_{N\to\infty} \int_{|x|+|y|\leq N} \overline{\psi}^{\pm}(k,x,y)\varphi(x,y)\, \frac{dx\, dy}{\mu_0(y)}, \tag{1.35}$$

$$(F^{0}\varphi)(k) = s - \lim_{N\to\infty} \int_{|x|+|y|\leq N} \overline{\psi}^{0}(k,x,y)\varphi(x,y)\, \frac{dx\, dy}{\mu_0(y)}, \tag{1.36}$$

and for $1 \leq j \leq Q$,

$$(F^{j}\varphi)(\overline{k}) = s - \lim_{N\to\infty} \int_{|x|+|y|\leq N} \overline{\psi}^{j}(\overline{k},x,y)\varphi(x,y)\, \frac{dx\, dy}{\mu_0(y)}, \tag{1.37}$$

exist respectively on the strong topology on $L^2(\mathbf{R}^{n+1}_{+})$, $L^2(\Omega_0)$, and $L^2(\mathbf{R}^n)$. The operators $F^{\pm}$, F^0, F^j, $1 \leq j \leq Q$, are partial isometric from $\mathcal{H}_{\mu_0}$ onto $L^2(\mathbf{R}^{n+1}_{+})$, $L^2(\Omega_0)$, and $L^2(\mathbf{R}^n)$. The operator

$$F_B\varphi = F^{+}\varphi \oplus F^{-}\varphi \bigoplus_{j=0}^{Q} F^{j}\varphi, \tag{1.38}$$

is unitary from $\mathcal{H}_{\mu_0}$ onto

$$L^2(\mathbf{R}_+^{n+1}) \oplus L^2(\mathbf{R}_+^{n+1}) \bigoplus_{j=0}^{Q} L^2(\mathbf{R}^n), \tag{1.39}$$

and for all $\varphi \in D(B)$

$$F_B B\, \varphi = (k^2 + q_-)F^+\varphi \oplus (k^2 + q_-)F^-\varphi \oplus (\overline{k}^2 - k_0^2 + q_-)F^0\varphi$$

$$\bigoplus_{j=1}^{Q} (\overline{k}^2 + \lambda_j)\, F^j\varphi. \tag{1.40}$$

In particular it follows that B is absolutely continuous and that $\sigma(B) = [\lambda_1, \infty)$.

In order to introduce the appropriate trace maps we define the following generalized eigenfunctions in polar coordinates

$$\psi^{\pm}(\sigma, \omega, x, y) = \sigma^{n/2}\, \psi^{\pm}(\sigma\omega, x, y), \tag{1.41}$$

for $\sigma > 0$, and $\omega \in S_{1,+}^n = S_1^n \cap \mathbf{R}_+^{n+1}$.

$$\psi^0(\sigma, k_0, \nu, x, y) = \sigma^{1/2}\, (g(\sigma, k_0))^{\frac{n-2}{2}}\, \psi^0(k_0, g(\sigma, k_0)\nu, x, y), \tag{1.42}$$

where

$$g(\sigma, k_0) = (\sigma^2 + k_0^2)^{1/2}, \tag{1.43}$$

for $\sigma > 0$, and $(k_0, \nu) \in S_1$, where

$$S_1 = (0, \sqrt{q_-}\,) \times S_1^{n-1}. \tag{1.44}$$

Moreover we define

$$\psi^j(\sigma, \nu, x, y) = \sigma^{\frac{n-1}{2}}\, \psi^j(\sigma\nu, x, y), \tag{1.45}$$

for $\sigma > 0$, $\quad \nu \in S_1^{n-1}$.

Then we have that

Lemma 1.1

For each $\sigma > 0$, and $s > 1/2$, there are trace maps, $T_{\pm}(\sigma)$, bounded from $L_s^2(\mathbf{R}^{n+1})$ into $L^2(S_{1,+}^n)$, such that for every $\varphi \in C_0^\infty(\mathbf{R}^{n+1})$

$$(T^{\pm}(\sigma)\varphi)(\omega) = \int \overline{\psi}^{\pm}(\sigma, \omega, x, y)\varphi(x, y)\, \frac{dx\, dy}{\mu_0(y)}. \tag{1.46}$$

Moreover the functions $\sigma \to T_{\pm}(\sigma)$ from $(0,\infty)$ into $\mathcal{B}(L_s^2(\mathbf{R}^{n+1}), L^2(S_{1,+}^n))$ are locally Hölder continuous with exponent γ that satisfies $\gamma < 1$, $\gamma < (s - 1/2)$.

Proof: Note that the coefficients α_+, $\beta_{\pm}$, $\gamma_{\pm}$, and $\delta_{\pm}$ in (1.10) and (1.11) are infinitely differentiable functions of ρ_- with bounded derivatives for $\rho_- \in [0,\infty)$, and that $\alpha_- = \rho_-\, g(\rho_-)$, where $g(\rho_-)$ is infinitely differentiable with bounded derivatives for $\rho_- \in [0,\infty)$. Then the lemma is proven as in cases (a) to (e) of Lemma 1.1 of Appendix 1.

Q. E. D.

Let us give to S_1 the product measure of the Lebesgue measure on $(0, \sqrt{q_1}\,)$, and the measure induced on S_1^{n-1} by Lebesgue measure on $\mathbf{R}^n$. By $L^2(S_1)$ we denote the Hilbert space of complex valued square integrable functions on S_1.

Lemma 1.2

For each $\sigma > 0$, $s > 1/2$, there is a trace map, $T^0(\sigma)$, bounded from $L_s^2(\mathbf{R}^{n+1})$ into $L^2(S_1)$, and such that for every $\varphi(x,y) \in C_0^{\infty}(\mathbf{R}^{n+1})$

$$(T^0(\sigma)\varphi)(k_0,\nu) = \int \overline{\psi}^0(k_0,\nu,x,y)\varphi(x,y)\,\frac{dx\,dy}{\mu_0(y)}. \tag{1.47}$$

Moreover the function $\sigma \to T^0(\sigma)$ from $(0,\infty)$ into $\mathcal{B}(L_s^2(\mathbf{R}^{n+1}),\ L^2(S_1))$ is Hölder continuous with exponent $\gamma < 1$, $\gamma < (s - 1/2)$.

Proof: For $\varphi \in C_0^{\infty}(\mathbf{R}^{n+1})$ we define $T^0(\sigma)$ by (1.47). Then

$$(T^0(\sigma)\varphi)(k_0,\nu) = \sqrt{\sigma}\,\left(g(\sigma,k_0)\right)^{\frac{n-2}{2}}\,\hat{f}_{k_0}(g(\sigma,k_0)\nu), \tag{1.48}$$

where

$$\hat{f}_{k_0}(\vec{k}) = \frac{1}{(2\pi)^{n/2}}\int e^{-i\vec{k}\cdot x}\,f_{k_0}(x)dx, \tag{1.49}$$

with

$$f_{k_0}(x) = \int \overline{\psi}_0(k_0,y)\varphi(x,y)\,\frac{dy}{\mu_0(y)}. \tag{1.50}$$

Note that there is a constant $\Delta > 0$, such that (see (1.19))

$$(\alpha_0^2 + \beta_0^2)^{1/2} > \Delta, \tag{1.51}$$

for $\rho_-^{\sim} \in (0, \sqrt{q_-}\,)$.

Then by the boundedness of the standard trace map $T_S(\rho)$ from$H_s(\mathbf{R}^n)$ into $L^2(S_1^{n-1})$ (see the proof of Lemma 1.1 in Appendix 1)

$$\begin{aligned}\left\|T_0(\sigma)\varphi\right\|^2_{L^2(S_1)} &= \int_0^{\sqrt{q_-}} dk_0\ \sigma\ g(\sigma,k_0)^{n-2} \\ &\int_{S_1^{n-1}} |T_S(g(\sigma,k_0))\hat{f}_{k_0}|^2\ d\nu \le \\ &\le C \int_0^{\sqrt{q_-}} dk_0 \int dx(1+|x|^2)^s\ |f_{k_0}(x)|^2 \le \\ &\le C \sum_{i=1}^{3} \int_0^{\sqrt{q_-}} dk_0 \int dx(1+|x|^2)^s\ |f_{k_0}^i(x)|^2,\end{aligned} \tag{1.52}$$

where for $k_0 = \rho_-^{\sim}$, and $\rho = (q_- - \rho_-^{\sim})^{1/2}$,

$$f^1_{\rho_-^{\sim}}(x) = \left(\frac{\rho_-^{\sim}}{\rho}\right)^{1/2} (\alpha_0+\beta_0)^{-1/2} \int_h^{\infty} (\alpha_0\cos(\rho y) + \beta_0 \sin(\rho y)) \cdot\ \varphi(x,y)\ \frac{1}{\mu_0(y)} dy, \tag{1.53}$$

$$f^2_{\rho_-^{\sim}}(x) = \left(\frac{\rho_-^{\sim}}{\rho}\right)^{1/2} (\alpha_0+\beta_0)^{-1/2} \int_0^{h} \left(\cos(\rho_0 y) + \frac{\mu_h\rho_-^{\sim}}{\mu_-\rho_0}\sin(\rho_0 y)\right) \cdot\ \varphi(x,y)\ \frac{1}{\mu_0(y)} dy, \tag{1.54}$$

$$f^3_{\rho_-^{\sim}}(x) = \left(\frac{\rho_-^{\sim}}{\rho}\right)^{1/2} (\alpha_0+\beta_0)^{-1/2} \int_{-\infty}^{0} e^{\rho_-^{\sim} y}\ \varphi(x,y)\ \frac{dy}{\mu_0(y)}. \tag{1.55}$$

Since the one dimensional Fourier transform is a unitary operator on $L^2(\mathbf{R})$

$$\int_0^{\sqrt{q_-}} dk_0\ |f^1_{k_0}(x)|^2 \le C \int_h^{\infty} |\varphi(x,y)|^2 dy. \tag{1.56}$$

By (1.54)

$$\int_0^{\sqrt{q-}} dk_0 \, |f^2_{k_0}(x)|^2 \leq C \int_0^h |\varphi(x,y)|^2 dy. \qquad (1.57)$$

Moreover by Schwarz inequality

$$\left| \int_{-\infty}^0 e^{\rho\tilde{-} y} \, \varphi(x,y) \, \frac{dy}{\mu_0(y)} \right|^2 \leq \frac{1}{2\rho\tilde{-}} \int_{-\infty}^0 |\varphi(x,y)|^2 dy. \qquad (1.58)$$

Then

$$\int_0^{\sqrt{q-}} dk_0 \, |f^2_{k_0}(x)|^2 \leq C \int_0^{\sqrt{q-}} \frac{d\rho\tilde{-}}{\rho} \int_{-\infty}^0 |\varphi(x,y)|^2 dy \leq$$

$$\leq C \int_{-\infty}^0 |\varphi(x,y)|^2 dy. \qquad (1.59)$$

By (1.52), and (1.56) to (1.59)

$$\left\| T^0(\sigma)\varphi \right\|_{L^2(S_1)} \leq C \left\| \varphi \right\|_{L^2_s(\mathbf{R}^{n+1})}. \qquad (1.60)$$

Since $T_S(\sigma)$ is locally Hölder continuous with exponent $\gamma < 1$, $\gamma < (s - 1/2)$, we similarly prove that $T^0(\sigma)$ is locally Hölder continuous as required by the Lemma.

Q. E. D.

Lemma 1.3

For each $\sigma > 0$, $s > 1/2$, and $1 \leq j \leq Q$, there is a trace map, $T^j(\sigma)$, bounded from $L^2_s(\mathbf{R}^{n+1})$ into $L^2(S_1^{n-1})$ such that for any $\varphi(x,y) \in C_0^\infty(\mathbf{R}^{n+1})$

$$(T^j(\sigma)\varphi)(\nu) = \int \psi^j(\sigma, \nu, x, y)\varphi(x,y) \, \frac{dxdy}{\mu_0(y)}. \qquad (1.61)$$

Moreover the functions $\sigma \to T^j(\sigma)$ from $(0,\infty)$ into $\mathcal{B}(L^2_s(\mathbf{R}^{n+1}), L^2(S_1^{n-1}))$ are locally Hölder continuous with exponent $\gamma < 1$, $\gamma < (s - 1/2)$.

Proof: We define $T^j(\sigma)$ by (1.61) for $\varphi(x,y) \in \mathbf{C}_0^\infty(\mathbf{R}^{n+1})$. Then

$$\int_{S_1^{n-1}} |T^j(\sigma)\varphi|^2 d\nu = \sigma^{n-1} \int_{S_1^{n-1}} |(T_S(\sigma)\, \hat{f})(\nu)|^2 d\nu, \tag{1.62}$$

where

$$\hat{f}(\bar{k}) = \frac{1}{(2\pi)^{n/2}} \int e^{-i\bar{k}\cdot x}\, f(x)dx, \tag{1.63}$$

with

$$f(x) = \int \overline{\psi}_j(y)\, \varphi(x,y)\, \frac{dy}{\mu_0(y)}. \tag{1.64}$$

Then since $T_S(\sigma)$ is bounded and $\|\psi_j\|_{\mathcal{N}_{\mu_0}} = 1$,

$$\int_{S_1^{n-1}} |T^j(\sigma)\varphi|^2 d\nu \leq C \int (1+|x|^2)^s\, |f(x)|^2 dx \leq$$

$$\leq C\, \|\varphi\|_{L_s^2(\mathbf{R}^{n+1})}. \tag{1.65}$$

Then $T^j(\sigma)$ is bounded from $L_s^2(\mathbf{R}^{n+1})$ into $L^2(S_1^{n-1})$.

Since $T_S(\sigma)$ is locally Hölder continuous we prove in a similar way that $T^j(\sigma)$ is locally Hölder continuous with exponent $\gamma < 1,\ \gamma < (s-1/2)$.

Q. E. D.

We define the operators

$$C^{\pm}(\sigma) = T^{\pm}(\sigma)\eta_{-s}, \tag{1.66}$$

$$C^j(\sigma) = T^j(\sigma)\eta_{-s}, \tag{1.67}$$

where $\sigma > 0,\ j = 0,1,2,\cdots,Q,\quad s > 1/2$, and $\eta_s(x,y) = (1+|x|^2+y^2)^{s/2}$.

For $z \in \mathbf{C} \setminus [\lambda_1, \infty)$ we denote by

$$R_B(z) = (B-z)^{-1}, \tag{1.68}$$

the resolvent of B.

For any $\lambda > q_-$, let a, b satisfy $q_- < a < \lambda < b$, and let us denote by I_λ the interval $I_\lambda = \left[\sqrt{a-q_-},\ \sqrt{b-q_-}\right]$. Then by (1.40) and functional calculus for any $s > 1/2$, and $\epsilon > 0$,

$$\eta_{-s}\, R_B(\lambda\, \pm\, i\epsilon)\eta_{-s} = \int\limits_{I_\lambda} \frac{1}{\sigma^2 + q_- - \lambda \mp i\epsilon} \Big[C^{+*}(\sigma)C^{+}(\sigma) +$$

$$+ C^{-*}(\sigma)C^{-}(\sigma) + C^{0*}(\sigma)C^{0}(\sigma)\Big]d\,\sigma + \sum_{j=1}^{Q} \int\limits_{I_j} \frac{1}{\sigma^2 + \lambda_j - \lambda \mp i\epsilon}$$

$$C^{j*}(\sigma)C^{j}(\sigma)d\,\sigma + \eta_{-s}\, R_B(\lambda\, \pm\, i\epsilon)\; E_B([a,b]^{\sim})\eta_{-s}, \tag{1.69}$$

where $I_j = \left[\sqrt{a - \lambda_j},\ \sqrt{b - \lambda_j}\,\right]$, $E_B(\,\cdot\,)$ denotes the family of spectral projectors of B and $[a,b]^{\sim}$ is the complement in $\mathbf{R}$ of $[a,b]$.

Then by (1.69) (see Musckhelishvili 1953, Privalov 1956, and Kuroda 1980, Theorem 5 and Remark 6 in section 4.1) the following limits

$$R_B(\lambda\, \pm\, i0) = \lim_{\epsilon\downarrow 0} R_B(\lambda\, \pm\, i\epsilon), \tag{1.70}$$

exist in the uniform operator topology on $\mathcal{B}(L^2_s(\mathbf{R}^{n+1}),\ L^2_{-s}(\mathbf{R}^{n+1}))$, $s > 1/2$, uniformly for λ in compact sets of (q_-, ∞), and

$$\eta_{-s}\, R_B(\lambda\, \pm\, i0)\eta_{-s} = P.V. \int\limits_{I_\lambda} \frac{1}{\sigma^2 + q_- - \lambda} \Big[C^{+*}(\sigma)C^{+}(\sigma) +$$

$$+ C^{-*}(\sigma)C^{-}(\sigma) + C^{0*}(\sigma)C^{0}(\sigma)\Big]d\,\sigma +$$

$$+ \sum_{j=1}^{Q} P.V. \int\limits_{I_j} \frac{1}{\sigma^2 + \lambda_j - \lambda}\, C^{j*}(\sigma)C^{j}(\sigma)d\,\sigma\, \pm$$

$$\pm\, i\pi\, \Big[C^{+*}(\lambda^{\sim})C^{+}(\lambda^{\sim}) + C^{-*}(\lambda^{\sim})C^{-}(\lambda^{\sim}) +$$

$$+ C^{0*}(\lambda^{\sim})C^{0}(\lambda^{\sim}) + \sum_{j=1}^{Q} C^{j*}(\lambda_j^{\sim})C^{j}(\lambda_j^{\sim})\Big] + \eta_{-s}\, R_B(\lambda)\; E_B([a,b]^{\sim})\eta_{-s}, \tag{1.71}$$

where $\lambda^{\sim} = (\lambda - q_-)^{1/2}$, and $\lambda_j^{\sim} = (\lambda - \lambda_j)^{1/2}$. $P.V.$ stands for the principal value of the integrals.

Let us denote

$$S_2 = S^n_{1,+}\ \oplus\ S^n_{1,+}\ \oplus\ S_1. \tag{1.72}$$

For $\varphi = \varphi_+(\sigma)\ \oplus\ \varphi_-(\sigma)\ \oplus\ \varphi_0(\sigma) \in L^2(I_\lambda,\ L^2(S_2))$ we define

$$M_0(\varphi) = \int\limits_{I_\lambda} \Big[C^{+*}(\sigma)\varphi_+(\sigma) + C^{-*}(\sigma)\varphi_-(\sigma) + C^{0*}(\sigma)\varphi_0(\sigma)\,\Big]d\sigma. \tag{1.73}$$

By (1.40)

$$M_0(\varphi) = \eta_{-s}\ E_B([a,b])\ F_B^*\ \varphi, \tag{1.74}$$

where we extended φ to $(\sqrt{q_-}, \infty) \setminus [a,b]$ by zero. Then

$$\left\| M_0(\varphi) \right\|_{L^2_s(\mathbf{R}^{n+1})} \le C \left\| \varphi \right\|_{L^2(I_\lambda,\ L^2(S_2))}, \tag{1.75}$$

and $M_0 \in \mathcal{B}\Big(L^2(I_\lambda,\ L^2(S_2)), L^2_s(\mathbf{R}^{n+1})\Big)$. Clearly $M_0 \in \mathcal{B}\Big(L^1(I_\lambda, L^2(S_2)), L^2(\mathbf{R}^{n+1})\Big)$. Then by interpolation (see Lions and Magenes 1972, and Reed and Simon 1975, Appendix to Section IX.4) $M_0 \in \mathcal{B}\Big(L^p(I_\lambda,\ L^2(S_2)), L^2_{\epsilon_p s}(\mathbf{R}^{n+1})\Big)$, where $\epsilon_p = 2(1 - \frac{1}{p})$, $1 \le p \le 2$.

For $1 \le j \le Q$, we define the operators M_j as

$$M_j(\varphi) = \int_{I_j} C^{j*}(\sigma)\varphi(\sigma)d\sigma. \tag{1.76}$$

We prove as above that $M_j \in \mathcal{B}\Big(L^p(I_j,\ L^2(S_1^{n-1})),\ L^2_{\epsilon_p s}(\mathbf{R}^{n+1})\Big)$, where $\epsilon_p = 2(1 - \frac{1}{p})$, $1 \le p \le 2$.

<u>**Lemma 1.4**</u>

For any $z \in \rho(B)$, $R_B(z)$ is a bounded operator from $L^2_s(\mathbf{R}^{n+1})$ into $L^2_s(\mathbf{R}^{n+1})$, for any $s \ge 0$.

<u>**Proof**</u>: The Lemma is proven as in the proof of Lemma 4.1 in Section 4 of Chapter 1.

Q. E. D.

We need in Section 5 the limiting absorption principle for the operator in $L^2(\mathbf{R})$

$$h_1 = -\frac{d^2}{dy^2} + q_1(y), \tag{1.77}$$

$$D(h_1) = H_2(\mathbf{R}), \tag{1.78}$$

where $q_1(y)$ is a real valued, bounded and measurable function. Moreover

$$|q_1(y)| \le C\ (1+y)^{-1-\epsilon}, \quad y > 0, \tag{1.79}$$

$$|q_1(y) - q_-| \leq C\ (1+|y|)^{-1-\epsilon}, \quad y < 0, \tag{1.80}$$

for some positive constants C, ϵ, and where $q_- \geq 0$. Note that with h as in (1.6) with $\mu_0(y) \equiv 1$,

$$h_1 = h + q_3(y), \tag{1.81}$$

where

$$q_3(y) = q_1(y) - q(y). \tag{1.82}$$

Note that

$$|q_3(y)| \leq C\ (1+|y|)^{-1-\epsilon}, \tag{1.83}$$

for some constants $C, \epsilon > 0$.

We easily prove as in Lemma 1.1 that for each $\sigma > 0,\ s > 1/2$, there are trace maps, $T_1^{\pm}(\sigma)$, bounded from $L_s^2(\mathbf{R})$ into $\mathbf{C}$, such that for each $\varphi(y) \in C_0^{\infty}(\mathbf{R})$

$$(T_1^{\pm}(\sigma)\varphi) = \int \psi_{\pm}(\sigma, y)\varphi(y)\ \frac{dy}{\mu_0(y)}. \tag{1.84}$$

Moreover the functions $\sigma \to T_1^{\pm}(\sigma)$ from $(0, \infty)$ into $\mathcal{B}(L_s^2(\mathbf{R}),\ \mathbf{C})$ are locally Hölder continuous with exponent $\gamma < 1,\ \gamma < (s - 1/2)$.

For $z \in \rho(h)$ denote

$$r_h(z) = (h - z)^{-1}. \tag{1.85}$$

Then by (1.24) for $\lambda > q_-, \quad \epsilon > 0$

$$v_{-s} r_h(\lambda \pm i\epsilon) v_{-s} = \int_{I_\lambda} \frac{1}{\sigma^2 + q_- - \lambda \mp i\epsilon}$$

$$\cdot \left[C_1^{+*}(\sigma)C_1^{+}(\sigma) + C_1^{-*}(\sigma)C_1^{-}(\sigma)\right] d\sigma + v_{-s} r_h(\lambda \pm i\epsilon) E_h([a,b]^{\sim}) v_{-s}, \tag{1.86}$$

where I_λ is as in (1.69),

$$C_1^{\pm}(\sigma) = T_1^{\pm}(\sigma)\ v_{-s}, \tag{1.87}$$

and for $s \in \mathbf{R}$

$$v_s(y) = (1 + |y|^2)^{s/2}, \tag{1.88}$$

and where $E_h(\ \cdot\ )$ denotes the spectral family of h.

Then as in the proof of Theorem 2.4 in Section 2 of Chapter 1, the limits

$$r_h(\lambda \pm i0) = \lim_{\epsilon \downarrow 0} r_h(\lambda \pm i\epsilon), \tag{1.89}$$

exist in the uniform operator topology in $\mathcal{B}(L^2_s(\mathbf{R}),\ H_{2,-s}(\mathbf{R}))$, $\quad s > 1/2$, uniformly for λ in compact sets of (q_-, ∞), and the functions

$$r_{h,\pm}(z) = \begin{cases} r_h(z), & Im\ z \neq 0, \\ r_h(z\ \pm\ i0), & z \in (q_-, \infty), \end{cases} \tag{1.90}$$

defined for $z \in \mathbf{C}^{\pm} \cup (q_-, \infty)$, are analytic for $Im\ z \neq 0$ and locally Hölder continuous for $z \in (q_-, \infty)$ with exponent $\gamma < 1,\ \gamma < (s - 1/2)$.

Moreover

$$r_h(\lambda\ \pm\ i0) = P.V. \int_{I_\lambda} \frac{1}{\sigma^2 + q_- - \lambda} \Big[C_1^{+*}(\sigma) C_1^+(\sigma) + C_1^{-*}(\sigma) C_1^-(\sigma) \Big] d\sigma +$$

$$\pm\ i\ \pi \left[C_1^{+*}(\tilde{\lambda}) C_1^+(\tilde{\lambda}) + C_1^{-*}(\tilde{\lambda}) C_1^-(\tilde{\lambda}) \right] + v_{-s} r_h(\lambda) E_h([a,b]^{\sim})\ v_{-s}, \tag{1.91}$$

where $\tilde{\lambda}$ is as in (1.71).

For $z \in \rho(h_1)$ denote

$$r_1(z) = (h_1 - z)^{-1}. \tag{1.92}$$

Note that for $z \in \rho(h_1) \cap \rho(h)$

$$r_1(z) = r_h(z)\ (1 + q_3\ r_h\ (z))^{-1}. \tag{1.93}$$

We prove as in Lemma 6.1 of Section 6 in Chapter 1 that

$$(1 + q_3\ r_{h,\pm}(z)), \tag{1.94}$$

are invertible on $L^2_s(\mathbf{R})$, $\quad s = \frac{1+\epsilon}{2}$, for $z \in \mathbf{C}^{\pm} \cup (q_-, \infty)$ (note that h_1 has no positive eigenvalues. See Eastham and Kalf 1982). Then for $\lambda > q_+$ the limits

$$r_1(\lambda\ \pm\ i0) = \lim_{\epsilon \downarrow 0} r_1(\lambda\ \pm\ i\epsilon), \tag{1.95}$$

exist in the uniform operator topology in $\mathcal{B}(L^2_s(\mathbf{R}),\ H_{2,-s}(\mathbf{R}))$, $\quad s = \frac{1+\epsilon}{2}$, uniformly for λ in compact sets of (q_-, ∞), and

$$r_1(\lambda\ \pm\ i0) = r_{h,\pm}(\lambda)\ (1 + q_3\ r_{h,\pm}(\lambda))^{-1}. \tag{1.96}$$

Moreover the functions

$$r_{1,\pm}(z) = \begin{cases} r(z), & Im\ z \neq 0, \\ r(z\ \pm\ i0), & z \in (q_-, \infty), \end{cases} \tag{1.97}$$

defined for $z \in \mathbf{C}^{\pm} \cup (q_-, \infty)$ are analytic for $Im\ z \neq 0$, and locally Hölder continuous for $z \in (q_-, \infty)$ with exponent $\gamma < 1,\ \gamma < (s - 1/2)$.

Notes

Chapter 2

The limiting absorption principle for the unperturbed acoustic propagator in Theorem 2.4 is due to Weder 1988, however a complete proof of the existence and local Hölder continuity of the trace maps (Lemmas 2.1, and 2.2) is published here for the first time in Appendix 1.

Theorem 5.1 is a generalization of a result on the absence of positive eigenvalues of the perturbed acoustic propagator published in Weder 1988b.

The complementary result on the absence of positive eigenvalues given in Theorem 5.4 is due to Weder 1988, where the Mourre estimate is proven. However the argument starting with the Mourre estimate given in the proof of Theorem 5.4 is published here for the first time.

The method of the limiting absorption principle was first studied by Eidus 1962.

Lemma 6.1 and Theorem 6.2 in the limiting absorption principle for the perturbed acoustic propagator are due to Weder 1988. They are a generalization of results of Agmon 1975, and Kuroda 1973, and 1980, for operators of Schrödinger type.

One of the new problems posed by the proof of the limiting absorption principle in the case of perturbed stratified media comes from the fact that the properly normalized generalized eigenfunctions of the unperturbed acoustic propagator are not smooth in the spectral parameter at thresholds (cut off frequencies). The method of solution at thresholds presented in Lemma 6.1, namely to change the problem into that of a related operator whose generalized eigenfunctions have better smoothness properties, appears to be new and it has independent interest.

It is an interesting fact that in the case of closed wave guides the situation is quite different. It has recently been proved by Werner 1984, 1985, 1987, and 1987b, and Morgenröther, Sindelfingen, and Werner 1987, that, for example, for the wave equation on the strip $x \in \mathbf{R}^n$, $0 < y < \pi$, $n = 1, 2$, with Neumann or Dirichlet boundary condition at $y = 0$, and $y = \pi$, neither the limiting absorption nor the limit amplitude principles are true, and that there are resonances at the thresholds $\lambda = 1, 2, 3, \cdots$.

The limiting absorption principle for the perturbed acoustic propagator was first proven in the case of all space $\mathbf{R}^{n+1}$, away from thresholds and under the decay condition

$$|c(x, y) - c_0(y)| \leq C\ (1 + |x|)^{-1-\delta_1}\ (1 + |y|)^{-1-\delta_2}, \tag{1}$$

$\delta_1,\ \delta_2 > 0$, and between weighted spaces with weight

$$\eta_{s_1, s_2}(x, y) = (1 + |x|^2)^{s_1/2}\ (1 + |y|^2)^{s_2/2}, \tag{2}$$

in Weder 1985, and Dermenjian and Guillot 1986, using different techniques.

It was proven constructively for transmission problems and exterior domains, for general boundary conditions satisfying the local compactness assumption, away from thresholds, under the decay condition (1) and with the weights (2) in Weder 1986. In Dermenjian and Guillot 1986 it was proven for exterior domains with Neumann and Dirichlet boundary conditions under the assumption that $c(x, y) - c_0(y)$ is of compact support, with the weights (2), away from thresholds and using a non constructive technique. They do not prove the Hölder continuity of the extended resolvents.

The results in the generalized Fourier maps for the perturbed acoustic propagator and on the acoustic scattering theory for transmission problems and exterior domains that we present in Section 7 are published here for the first time.

Chapter 3

The results in Section 1 on the unperturbed electromagnetic propagator that allow us to obtain our Theorem 1.3 in generalized eigenfunctions expansions are published here for the first time.

Our coerciveness result for the unperturbed electromagnetic propagator (Lemma 2.4) is first published here.

Theorem 2.5 on the limiting absorption principle for the unperturbed electromagnetic propagator was stated in Weder 1988c. However the proof is given here for the first time.

Theorem 3.7 in the limiting absorption principle for the perturbed electromagnetic propagator is due to Weder 1988c. The method of proof is a generalization of the one used in the acoustic case in Theorem 6.2 in Chapter 2.

The results in Section 4 on the generalized Fourier maps and electromagnetic scattering theory, as well as those in Appendix 2 are due to Weder 1988c.

A related problem, namely the limiting absorption principle for elastic waves in a perturbed isotropic half space has been studied in Dermenjian and Guillot 1988.

References

Adams, R.A., 1975. *Sobolev Spaces.* Academic Press. New York, San Francisco, London.

Agmon, S., 1975. *Spectral Properties of Schrödinger Operators and Scattering Theory.* Analli della Scuola Norm. Sup. di Pisa cl. Sci., 2, 151-218.

Brekhovskikh, L.M., 1980. *Waves in Layered Media.* Second Edition. Academic Press. New York, London, Toronto, Sydney, San Francisco.

Combes, J.M., and Weder, R., 1981. *New Criterion for Existence and Completeness of Wave Operators and Applications to Scattering by Unbounded Obstacles.* Comm. in Part. Diff. Equations, 6, (11), 1179-1223.

Dermenjian, Y., and Guillot, J.C., 1986. *Théorie Spectrale de la Propagation des Ondes Acoustiques dans un Milieu Stratifié Perturbé.* J. of Diff. Equ., 62, N° 3, 357-409.

Dermenjian Y., and Guillot, J.C., 1988 *Scattering of Elastic Waves in a Perturbed Isotropic Half Space with a Free Boundary. The Limiting Absorption Principle.* Math. Meth. in the Appl. Sci., 10, 87-124.

Dixmier, J., 1969. *Les Algèbres d'Opèrateurs dans l'Espace Hilbertien.* Gauthier Villars. Paris.

Dunford, N., and Schwartz, J., 1963. *Linear Operators.* Part II. Spectral Theory Interscience Publishers. New York, London.

Eidus, 1962. *The Principle of Limiting Absorption.* Mat. Sb., 57, 13-44. And A.M.S. Transl. (2), 47, (1965) 157-191.

Eastham, M.S.P., and Kalf, H., 1982. *Schrödinger Type Operators with Continuous Spectrum.* Research Notes in Mathematics 65. Pitman. Boston, London, Mebourne.

Froese, R., and Herbst, I., 1982. *Exponential Bounds and Absence of Positive Eigenvalues for N-Body Schrödinger Operators.* Comm. Math. Phys., 87, 429-447.

Kato, T., 1967. *Scattering Theory with Two Hilbert Spaces.* J. Functional Analysis, 1, 342-369.

Kato, T., 1976. *Perturbation Theory for Linear Operators.* Springer-Verlag. Berlin, Heidelberg, New York.

Kuroda, S.T., 1973. *Scattering Theory for Differential Operators I..* J. Math. Soc., Japan, 25, N° 1, 75-104.

Kuroda, S.T., 1980. *An Introduction to Scattering Theory.* Lecture Note Series N° 51. Matematisk Institut. Aarhus Universitet.

Lions, J.L., and Magenes, E., 1972. *Non Homogeneous Boundary Value Problems and Applications.* Vol. I. Springer-Verlag. Berlin, Heidelberg, New York.

Marcuse, D., 1974. *Theory of Dielectric Optical Wave Guides.* Academic Press. London, New York, San Francisco.

Morgenröther, K., Sindelfingen, and Werner, P. *Resonances and Standing Waves.* Math. Meth. in the Appl. Sci., 9, 105-126.

Mourre, E., 1981. *Absence of Singular Continuous Spectrum for Certain Selfadjoint Operators.* Comm. Math. Phys., 78, 391-408.

Muskhelishvili, N.I., 1953. *Singular Integral Equations.* P. Noordhoff. Groningen.

Pekeris, C.L., 1948. *Theory of Propagation of Explosive Sound in Shallow Water.* Geol. Soc. Am. Memoir 27.

Privalov, I. I., 1956. *Randeigenschaften Analytischer Funktionen.* VEB Deutcher Verlag der Wissenschaften. Berlin.

Reed, M., and Simon, B., 1972. *Methods of Modern Mathematica Physics.* Vol. I. Academic Press. New York, San Francisco, London.

Reed, M., and Simon, B., 1975. *Methods of Modern Mathematical Physics.* Vol. II. Academic Press. New York, San Francisco, London.

Reed, M., and Simon, B., 1978. *Methods of Modern Mathematical Physics.* Vol. IV. Academic Press. New York, San Francisco, London.

Reed, M., and Simon, B., 1979. *Methods of Modern Mathematical Physics.* Vol. III. Academic Press. New York, San Francisco, London.

Royden, H.L., 1968. *Real Analysis.* MacMillan Publishing Co. New York.

Schechter, M., 1971. *Spectra of Partial Differential Operators.* Elsevier. North Holland, Amsterdam.

Tolstoy, I., and Clay, C.S., 1966. *Ocean Acoustic.* McGraw-Hill. New York.

Weder, R., 1984. *Scattering Theory for High Order Operators in Domains with Infinite Boundary.* Journal of Functional Analysis, 57, N° 2, 207-231.

Weder, R., 1984b. *Spectral Analysis of Strongly Propagative Systems.* J. für die reine und angew Mathematik, 354, 95-122.

Weder, R., 1985. *Spectral and Scattering Theory in Perturbed Stratified Fluids.* J. Math. Pures et Appl., 64, 149-173.

Weder, R., 1986. *Spectral and Scattering Theory in Perturbed Stratified Fluids II. Transmission problems and Exterior domains.* J. Diff. Equ., 64, N° 1, 109-131.

Weder, R., 1988. *The Limiting Absorption Principle at Thresholds.* J. Math. Pures et Appl., 67, 313-338.

Weder, R., 1988b. *Absence of Eigenvalues of the Acoustic Propagator in Deformed Wave Guides.* Rocky Mount. J. of Math., 18, N° 2, 495-503.

Weder, R., 1988c. *Spectral and Scattering Theory in Deformed Optical Wave Guides.* J. für die reine angew. Math., 390, 130-169.

Werner, P., 1984. *Ein Resonanzphänomen in der Theorie Akustischer und Electromagnetischer Wellen.* Math. Meth. Appl. Sci., 6, 104-128.

Werner, P., 1985. *Zur Asymptotik der Wellengleichung und der Wärmeleitungsgleichung in Zweidimensionalen Außenräumen.* Math. Meth. Appl. Sci., 7, 170-201.

Werner, P., 1987. *Resonance Phenomena in Cylindrical Waves Guides.* J. Math. Anal. and Appl., 121, N° 1, 173-214.

Werner, P., 1987b. *Aperiodic Electromagnetic Waves in Cylindrical Wave Guides.* J. Math. Anal. and Appl., 121, 215-272.

Wilcox, C., 1975. *Scattering Theory for the D'Alembert Equation in Exterior Domains.* Lecture Notes in Mathematics, Vol. 442. Springer-Verlag. New York, Berlin.

Wilcox, C., 1976. *Spectral Analysis of the Pekeris Operator in the Theory of Acoustic Wave Propagation in Shallow Water.* Arch. Rat. Mech. An., 60, N° 3, 259-300.

Wilcox, C., 1976b. *Transient Electromagnetic Wave Propagation in a Dielectric Wave Guide.* Instituto Nazionale di Alta Matematica. Symposia Matematica XVIII, 239-277.

Wilcox, C., 1984. *Sound Propagation in Stratified Fluids.* Applied Mathematical Sciences. 50. Springer- Verlag. New York, Berlin, Heidelberg.

Index

Applied Mathematical Sciences

52. *Chipot:* Variational Inequalities and Flow in Porous Media.
53. *Majda:* Compressible Fluid Flow and Systems of Conservation Laws in Several Space Variables.
54. *Wasow:* Linear Turning Point Theory.
55. *Yosida:* Operational Calculus: A Theory of Hyperfunctions.
56. *Chang/Howes:* Nonlinear Singular Perturbation Phenomena: Theory and Applications.
57. *Reinhardt:* Analysis of Approximation Methods for Differential and Integral Equations.
58. *Dwoyer/Hussaini/Voigt (eds.):* Theoretical Approaches to Turbulence.
59. *Sanders/Verhulst:* Averaging Methods in Nonlinear Dynamical Systems.
60. *Ghil/Childress:* Topics in Geophysical Dynamics: Atmospheric Dynamics, Dynamo Theory and Climate Dynamics.
61. *Sattinger/Weaver:* Lie Groups and Algebras with Applications to Physics, Geometry, and Mechanics.
62. *LaSalle:* The Stability and Control of Discrete Processes.
63. *Grasman:* Asymptotic Methods of Relaxation Oscillations and Applications.
64. *Hsu:* Cell-to-Cell Mapping: A Method of Global Analysis for Nonlinear Systems.
65. *Rand/Armbruster:* Perturbation Methods, Bifurcation Theory and Computer Algebra.
66. *Hlaváček/Haslinger/Nečas/Lovíšek:* Solution of Variational Inequalities in Mechanics.
67. *Cercignani:* The Boltzmann Equation and Its Applications.
68. *Temam:* Infinite Dimensional Dynamical System in Mechanics and Physics.
69. *Golubitsky/Stewart/Schaeffer:* Singularities and Groups in Bifurcation Theory, Vol. II.
70. *Constantin/Foias/Nicolaenko/Temam:* Integral Manifolds and Inertial Manifolds for Dissipative Partial Differential Equations.
71. *Catlin:* Estimation, Control, and the Discrete Kalman Filter.
72. *Lochak/Meunier:* Multiphase Averaging for Classical Systems.
73. *Wiggins:* Global Bifurcations and Chaos.
74. *Mawhin/Willem:* Critical Point Theory and Hamiltonian Systems.
75. *Abraham/Marsden/Ratiu:* Manifolds, Tensor Analysis, and Applications, 2nd ed.
76. *Lagerstrom:* Matched Asymptotic Expansions: Ideas and Techniques.
77. *Aldous:* Probability Approximations via the Poisson Clumping Heuristic.
78. *Dacorogna:* Direct Methods in the Calculus of Variations.
79. *Hernández-Lerma:* Adaptive Markov Control Processes.
80. *Lawden:* Elliptic Functions and Applications.
81. *Bluman/Kumei:* Symmetries and Differential Equations.
82. *Kress:* Linear Integral Equations.
83. *Bebernes/Eberly:* Mathematical Problems from Combustion Theory.
84. *Joseph:* Fluid Dynamics of Viscoelastic Liquids.
85. *Yang:* Wave Packets and Their Bifurcations in Geophysical Fluid Dynamics.
86. *Dendrinos/Sonis:* Chaos and Socio-Spatial Dynamics.
87. *Weder:* Spectral and Scattering Theory for Wave Propagation in Perturbed Stratified Media.

9 780387 973579